Are We Alone?

The Stanley Kubrick Extraterrestrial-Intelligence Interviews

Are We Alone?

The Stanley Kubrick Extraterrestrial-Intelligence Interviews

Twenty-one leading scientists speculate on life in the universe and the impact its discovery would have on us

Edited with an Introduction, Notes, etc.
by Anthony Frewin

Preface by Sir Arthur C. Clarke

Afterword by Frederick I. Ordway III

Ashgrove Publishing
London

It's reasonable to assume that there must be, in fact, countless billions of such planets where biological life has arisen, and the odds of some proportion of such life developing intelligence are high.

STANLEY KUBRICK (1968)

Contents

7 Introduction to the 2018 Edition
11 Preface by Sir Arthur C. Clarke
13 Introduction
35 Acknowledgements

THE INTERVIEWS
37 Isaac Asimov
45 Jeremy Bernstein / Addendum (2003) – 52
53 Frank D. Drake
63 Frederick C. Durant III / Looking Back (2003) – 69
71 Freeman J. Dyson
79 Gerald Feinberg
89 Constantine D. J. Generales Jr
99 Irving John Good / Footnotes to the Interview (2004) – 115
117 Gerald S. Hawkins / Forty Years On (2003) – 124
127 Francis J. Heyden SJ
135 Norman Lamm
145 Sir Bernard Lovell / Addendum (2003) – 151
153 Margaret Mead
165 Marvin Minsky
175 Ormond G. Mitchell
181 Philip Morrison / On Distant Earths (2003) – 191
193 A. I. Oparin
197 Harlow Shapley
209 S. M. Siegel
221 B. F. Skinner
229 Fred L. Whipple

APPENDICES
237 I: A Theological Take: God in *2001: A Space Odyssey* by Dr. John Braverman SJ
241 II: Other Scientists Who Were Approached
243 III: The Roger Caras Letter
245 IV: Notes on the Interviewees

249 Afterword by Frederick I. Ordway III
255 Bibliography & Web Resources
263 L'Envoi

Arthur C. Clarke

Introduction to the 2018 Edition

Anthony Frewin

Since this book was first published in 2005 we haven't been contacted by extraterrestrial entities and neither have we discovered life of any form on other planets, yet there is a greater buoyancy and optimism abroad in this area. Why? One reason is the discovery by NASA's Kepler space telescope of well over 3,000 exoplanets, and those are just the ones it has observed. Life must exist on some of them. As individuals we crave company, and as humanity we crave cosmic company. Surely we are not alone in this vast and probably infinite universe? Are we as common as dirt or are we unique? As Arthur C. Clarke noted: 'Two possibilities exist: either we're alone in the universe, or we're not. Both are equally terrifying.'

Well, the year 2001 came and went and now we are in 2018 which is the fiftieth anniversary of the release of *2001: A Space Odyssey*, a fact hard to believe by those of us still about who worked on the film. Yes, fifty years, but it is comforting to know that *2001* still has the import and pull now that it had when it was first released. A year or two back I said a few words about the film at an open air screening at the Jodrell Bank radio telescope. The audience was just a little shy of 500 people, and it seemed to me that around 80% of them hadn't even been born when the film came out. *2001* lives on.

Since the original publication of this book we have lost a number of contributors: Fred Durant passed away in 2015 at the grand age of 98, I. J. Good went in 2009, Sir Bernard Lovell in 2012, Marvin Minsky in 2016, and Fred Whipple in 2004 (too late in that year to be noted in the first edition). The sole surviving interviewee is that illustrious professor emeritus at the Institute for Advanced Study in Princeton, Freeman Dyson (born 1923).

And, yes, two *2001* alumni: Arthur C. Clarke departed this world in 2008 while Fred Ordway followed him in 2014. *Requiescat.*

I have left the bibliographies and other references at the end as they appeared in 2005, and here below note only those titles published since then that have struck me as being of some significance.

The listings below are not intended to be exhaustive, rather they merely reflect my own desultory reading.

EXTRATERRESTRIAL INTELLIGENCE: GENERAL WORKS.

BERTKA, CONSTANCE M. (Ed.). *Exploring the Origin, Extent, and Future of Life: Philosophical, Ethical and Theological Perspectives.* Cambridge: Cambridge University Press, 2009.

CIRKOVIC, MILAN M. *The Astrobiological Landscape: Philosophical Foundations of the Study of Cosmic Life.* Cambridge: Cambridge University Press, 2012.

DICK, STEVEN J. (Ed.). *The Impact of Discovering Life beyond Earth.* Cambridge: Cambridge University Press, 2015. A stimulating collection of essays.

DICK, STEVEN J. & STRICK, JAMES. *The Living Universe: NASA and the Development of Astrobiology.* New Brunswick, NJ: Rutgers University Press, 2005.

DOWD, MATTHEW F., & VAKOCH, DOUGLAS A. (Eds.). *The Drake Equation: Estimating the Prevalence of Extraterrestrial Life through the Ages.* Cambridge: Cambridge University Press, 2015. Some fourteen essays by divers hands. Foreword by Frank Drake.

KOLB, VERA M. *Astrobiology: An Evolutionary Approach.* Boca Raton, FL: CRC Press, 2015.

LINCOLN, DON. *Alien Universe: Extraterrestrial Life in Our Minds and in the Cosmos.* Baltimore: John Hopkins University Press, 2013.

SCOLES, SARAH. *Making Contact: Jill Tarter and the Search for Extraterrestrial Intelligence.* New York: Pegasus Books, 2017.

SHUCH, H. PAUL (Ed.). *Searching for Extraterrestrial Intelligence: SETI Past, Present, and Future.* New York: Springer, 2011. A rich collection of essays.

WEBB, STEPHEN. *If the Universe Is Teeming with Aliens… Where Is Everybody? Fifty Solutions to the Fermi Paradox and the Problem of Extraterrestrial Life.* New York: Springer, 2010. Where, indeed, are they? The answer must be here somewhere in the fifty solutions discussed herein. An essential work.

• • •

THE FILM: *2001*: *A SPACE ODYSSEY*

BENSON, MICHAEL. *Space Odyssey*: *Stanley Kubrick, Arthur C. Clarke, and the Making of a Masterpiece.* New York: Simon & Schuster, 2018. The best account of the making of the film and the personalities involved and written with a novelist's eye for the telling detail. Benson also discusses the ideas of the film and its genesis through biographical portraits of both SK and ACC.

BIZONY, PIERS. *The Making of Stanley Kubrick's 2001*: *A Space Odyssey.* Cologne: Taschen, 2014. An elegantly designed picture book let down somewhat by Bizony's journalistic text. The making of the film is far better discussed by Benson above.

KOLKER, ROBERT (Ed.). *Stanley Kubrick's 2001*: *A Space Odyssey*: *New Essays.* New York: Oxford University Press, 2006. Some nine essays including R. Barton Palmer's 'The Critical Reception and the Generation Gap' that is a very good account as to why kids liked the film and their parents didn't.

THE CINEMA OF STANLEY KUBRICK

BRODERICK, MICK. *Reconstructing Strangelove*: *Inside Stanley Kubrick's 'Nightmare Comedy'*. New York: Wallflower Press/Columbia University Press, 2017. A model of what a film study should be. Witty and incisive, and readable in a way that most 'film studies' aren't.

CASTLE, ALISON, HARLAN, JAN, & KUBRICK, CHRISTIANE. *The Stanley Kubrick Archives.* Cologne: Taschen 2008. Monumental. Should be on the shelf of every SK student and fan.

ALISON, CASTLE (Ed.). *Stanley Kubrick's Napoleon*: *The Greatest Movie Never Made.* Cologne: Taschen, 2011, 2017. The next best thing to the movie that was never made.

I shouldn't forget HAL and Artificial Intelligence (AI). A major work on the subject is by the American academic Nils J. Nilsson: *The Quest for Artificial Intelligence*: *A History of Ideas and Achievements* (Cambridge: Cambridge University Press, 2009).

Stanley Kubrick used to say, read one book on any subject and you'll be amongst the top 10% of world authorities on that subject. Read two and you'll be in the top 5%. Not always an easy book, but Nilsson will put you in that select 10%.

AI, fifty years ago, was felt in many quarters to be more or less just around the corner, but in the ensuing decades it has receded further and further in to the future. That initial optimism has evaporated in the face of our increased understanding of the human brain and a more realistic understanding of just what computers can and can't do. It has been suggested that you cannot program intelligence, that the intelligent computer must be achieved by other means, though what these are we do not yet know.

It is sad to report that the missing film of the following interviews has not come to light in the last thirteen years. The reels are out there somewhere. I'm beginning to think they might be in that warehouse at the end of *Raiders of the Lost Ark*.

Yes, fifty years since *2001* was released, and I'm sure that come its centenary in 2068 there will be hosannas throughout the land. It is indeed a film for the ages.

Preface

by Sir Arthur C. Clarke

GOING THROUGH these interviews has given me the sensation, in Yogi Berra's immortal phrase, of 'déjà vu all over again.' It has also made me feel quite sad, because so many of those involved – including Roger Caras himself, who became a very close friend – are no longer with us. And I am indeed sorry that Stanley never lived to see the year he made famous.

How this entire project started I described in 'Son of Dr. Strangelove' (republished in *Greetings, Carbon-based Bipeds!*) and *The Lost Worlds of 2001*. Briefly, it was Stanley's idea that in order for people to take the idea of space travel and the possibility of extraterrestrial life seriously, we should preface our hoped-for movie with a series of comments by distinguished contemporary (1966!) thinkers.

Fortunately this was dropped at a very early stage (Jeremy Bernstein once remarked: 'It was one of Stanley Kubrick's few really bad ideas') but the interviews remain, and provide a fascinating snapshot of thought on space travel, and our place in the universe, during the pre-Apollo decade.

One person who should have been in this volume was Carl Sagan, who I introduced to Stanley when we were starting work on the movie. Unfortunately, Carl (unlike the other interviewees in this volume) proposed an astronomical fee for his services, so Stanley asked me not to bring him back for an interview that had already been scheduled. I made my excuses to Carl, saying that Stanley was indisposed, and we went instead to the just-opened New York's World Fair.

During a sudden rainstorm we took refuge in an exhibition which screened a very well-made film on astronomy. Unfortunately, it had been made by some fundamentalist group, so it ended with a commercial for God, to which Carl took great exception. He protested to one of the cinema staff on the way out, which I thought was rather unfair, as the poor man probably had nothing to do with the production. However, having since seen how Hollywood executives' careers can make sudden nose-dives, I am not so sure of this. I have since wondered if our little expedition may have triggered Carl's book and the movie, *Contact*.

Another friend in this volume whom I miss badly is Isaac Asimov. We formed a mutual admiration society and enjoyed exchanging insults over the years. Perhaps the best-known one concerns a press-cutting I sent Isaac, recording how when an airliner was in trouble, and expected to crash, one of the passengers distracted himself by reading an ACC book. 'He should have been reading one of yours, Isaac,' I wrote, 'then he would have slept through the whole thing.' Isaac promptly replied: 'The reason he was reading one of yours was that if the plane did crash, death would have been a merciful relief.' (Score: Asimov 1, Clarke 1.)

It is astonishing and somewhat disappointing to see how many of the issues raised in this volume are still completely unresolved, especially the question of the existence of intelligent life elsewhere, and the possibility of visits in the remote past. I still hope that this may be settled in my lifetime.

Finally, here is a profound observation – not original, alas – with which I have been boring all my friends: 'The best proof that there is intelligent life in outer space is the fact that it hasn't come here.'

Sri Lanka, Christmas 2004

Introduction

by Anthony Frewin

IN WRITING THIS INTRODUCTION I feel rather like a warm-up comedian at the beginning of a TV show. Nobody has arrived here to see me, everyone is impatient for the main attraction, but nevertheless there is an important if minor function to be performed. So, I will be as brief and concise as possible and, as Ned Polsky once wrote, 'flat-footedly direct, and [therefore?] difficult for readers inclined to skim.'[1]

When Stanley Kubrick began pre-production of *2001: A Space Odyssey* in 1965 he had a problem. In fact, he had many problems, but there was one particular problem, initially, anyway, that vexed him greatly: the public perception of science fiction or futuristic films. It may be difficult for anyone under the age of 50 to appreciate the fact that prior to *2001* science fiction in the cinema was something for Saturday morning kids' showings and the 'B' film circuit. It was UFOs, flying saucers, little green men, 'Mars Attacks,' and Flash Gordon and Buck Rogers.

It was the cinema engendered by the fantasy pulp magazines of the 1930s and 1940s and produced by the *schlockmeisters* of Hollywood's Poverty Row.[2]

---

1 *Hustlers, Beats, and Others*, 'Expanded Edition' (New York: The Lyons Press, 1998), p. xvi. First published in 1967. The bracketed interpolation is mine.

2 I may have spoken too soon. Science fiction in the cinema has returned to its pre-*2001* state. There seems to be some inverse law at work here: as the special effects have got better and better the films have got more and more banal. The SFX in, say, *Independence Day* (1996) or *Lost in Space* (1998) or *The Matrix Reloaded* (2003) are beyond anything any of us could have imagined when we were making *2001*, yet the story and content of these representative films is juvenile. Several people have pointed to *Contact* (1997), the film based on the Carl Sagan novel, as having breached the barriers, but I would not agree. It has the ponderous faux seriousness that one used to associate with Stanley Kramer. There is room in the cinema for intelligent, well-crafted hokum – Ridley Scott's *Alien* (1979) and its progeny are good examples – but where is the movie with the ambition and import of *2001*? Andrei Tarkovsky's *Solaris* (1972) comes to mind, but what else?

True, there may have been one or two exceptions in the history of the cinema, but these were largely forgotten in the popular imagination.[3]

Breaking this perception down one came up with two issues in regard to *2001*: space travel and the idea of extraterrestrial life, the latter being one of the major themes of the film. The subjects were not taken seriously and could hardly be the basis for what is commonly known to the film industry's barkers and shills as a *major motion picture*.

SK felt that he needed primarily to establish that the question of extraterrestrial life was a legitimate subject for scientific investigation and the question *Are we alone?* was of the most profound importance to mankind. If he could do this, acceptance of the other issues would follow. Without doing this he feared that *2001* would be dismissed as just another romp in the galaxy, another space opera, the perfect accompaniment for popcorn.

SK's solution to this problem was simple. He would interview on film the most important scientists, theologians, astronomers and philosophers who had confronted the problem of extraterrestrial life and 'prefix' *2001: A Space Odyssey* with an introduction made up of edited clips from those interviews, much in the manner of Herman Melville collecting quotations on whales and whaling from various authorities at the beginning of *Moby Dick*. This was a direct and elegant way of setting the scene and letting the audience know that what was coming might be worthy of more than passing attention. To this purpose, SK carefully drew up a list of questions, researched those he wished to interview[4] and sent his associate, Roger Caras, hither and thither around the world in 1966 interviewing on 35mm black-and-white film twenty-one scientists and others that he felt could contribute to the production.[5]

---

3 Two films preceding *2001* that one might consider as sharing its grandeur and ambition are Fritz Lang's *Metropolis* (1927) and the Alexander Korda/William Cameron Menzies *Things to Come* (1936), with a screenplay by H.G. Wells from his novel. SK thought that *Metropolis*, despite its imaginative art direction and cinematic strengths was, essentially, silly. None of us demurred. However, neither I nor Arthur C. Clarke could convince SK of the merits of *Things to Come* which he considered politically simplistic and preachy, although he did admire the special effects. Interestingly, two of the *2001* special effects supervisors, the gentlemanly Tom Howard and Wally 'Where's Wally?' Veevers, both began their careers working on *Things to Come* under Ned Mann, the SFX chief. Christopher Frayling's *Things to Come* (London: British Film Institute, 1995) is a useful little study of the film that does not shirk its political implications.

And while Caras was interviewing them he could also ask questions about artificial intelligence, the future of computing, the origin of life and other matters that could contribute to an accurate portrayal of the year 2001 (SK was never one to miss an opportunity).

As the production of *2001* fast approached completion in the summer of 1968 SK realised that the film would have to stand on its own two feet, that the interviews might in fact be a confusing distraction for the audience and, besides, with the picture already running to 160 minutes, there just was not enough time to include them. So, the cans of 35mm film were consigned to the vaults.

In 2004 I proposed to Christiane Kubrick and Jan Harlan that the interviews should be issued on video and DVD, possibly in conjunction with a re-release of *2001*. Aside from their association with the film, the interviews had an intrinsic interest and a historic interest, presenting us with snapshots as it were of thinking on ETI (Extra-Terrestrial Intelligence), artificial intelligence, and other matters as it was in 1966.

Christiane and Jan agreed that it would be a good idea and I went in search of the twenty-one film cans (which I last saw on my desk at MGM Studios in 1968). They were not in SK's archives and store rooms and none of the motion picture labs we dealt with could locate them in their vaults, either in England or in the US. I looked everywhere except eBay. SK would not have junked them and he certainly would not have given them away. My own feeling is that they are in the vaults of one of the films labs, but mis-identified. One day someone will chance upon them. I only hope they have the nous to appreciate their importance.

What I did find, however, were the four foolscap ring-binders containing the typed transcripts of the interviews. Thus, if the footage was unavailable, the next best thing would be to publish those transcripts.

THE SCIENTISTS AND THE INTERVIEWS

The twenty-one scientists and others are here presented in alphabetical order. Underneath their names at the beginning of each interview are their credits

---

4 Principal input on names was from Fred Ordway, but Arthur C. Clarke and Jeremy Bernstein also contributed.

5 The letter of invitation was drafted by SK. It is reproduced in Appendix III.

as they were in 1966. This is what would have appeared on screen as a subtitle card had the original proposal been realised. As best I can remember, all the interviews were filmed at the location noted in the credit, including Caras going to Moscow to see Oparin, though this was not an interview. Oparin gave a prepared statement which would have been read by an actor had SK gone ahead with the prologue.

All of those interviewed were or have become major international figures in their fields and, to underscore their importance, several of them are included in David W. Swift's recent and much praised *SETI Pioneers: Scientists Talk About Their Search for Extraterrestrial Intelligence.* [6]

To take a number of names at random: Aleksandr Ivanovich Oparin (1894-1980), the most important twentieth-century scientist to research the origins of life, wrote what has been described as the first and principal modern appreciation of the problem. Harlow Shapley (1885-1972) was a great American astronomer, an eloquent writer, and someone to whom we owe our understanding of the size and shape of our galaxy. B. F. Skinner (1904-90) is widely known for his work on behavioural conditioning. Freeman J. Dyson (1923–) invented the Dyson sphere. Dr Margaret Mead (1901-78) was the author of *Coming of Age in Samoa* (1928) and other important contributions to anthropology. Frank D. Drake (1930–) is a pioneer of SETI and radio astronomy and the formulator of the Drake Equation (see below). Fred Whipple might well be considered *the* great astronomer of the twentieth century.

There is one scientist, however, who is conspicuously absent, and that is Carl Sagan, as Arthur C. Clarke notes in his preface. The story that has been doing the rounds for the last forty years was that Sagan asked for a fee of $10,000, virtually the entire budget for the prologue. In fact, the reality is even more surprising. After Sagan was approached he wrote back that since the film was a 'commercial enterprise' he 'would consider a fair remuneration to be 0.002 per cent of the gross receipts of the film per minute of actual appearance by me in the film.'[7] Further, Sagan also wanted editorial control. This was a little too much for SK. Caras replied by saying that 'It will not be possible for us to provide you with editorial control. The project is far too expensive and too many people are involved.' However, 'We could give an assurance that we would not knowingly distort anything you had to say. We cannot offer

---

6 Tucson: University of Arizona Press, 1990.

7 Letter from Carl Sagan, Harvard College Observatory, dated 10 March 1966, to Roger Caras, MGM Studios, Boreham Wood.

financial compensation for participation in the Prologue.'[8] Sagan, as they say in the vernacular, took a hike.[9] Others were approached and for various reasons could not appear. I have listed them in an appendix.

It is unnecessary for me to add a gloss to the interviews. The reader wishing to follow up on any of the themes discussed is referred to the bibliography at the end. I would just like to point to the optimism that was abroad in 1966. Both Marvin Minsky and Gerald Feinberg believed that by the year 2001 we would have computers comparable in intelligence to man's intelligence (HAL, in fact) while I. J. 'Jack' Good had no doubt that by then we would be in contact with extraterrestrial entities. Several of the interviewees also believed that the various space programmes would have continued in leaps and bounds and we would now be colonising our solar system. None of this has come to pass (but 'Tino' Generales accurately predicted the Internet).

There are inherent pitfalls in predicting the future and this was pithily expressed by the American writer, Bernard Wolfe, some fifty years ago at the end of a wonderful but now neglected novel that SK greatly admired:

> This book, then, is a rather bilious rib on 1950 – on what 1950 might have been like if it had been allowed to fulfil itself, if it had gone on being 1950,

---

8 Caras to Carl Sagan from MGM Studios, Boreham Wood, 15 March 1966.

9 Nothing about the proposed interview is mentioned in Keay Davidson's recent biography, *Carl Sagan: A Life* (New York: John Wiley, 1999). However, Davidson needs to be corrected on the non-representation of the alien entities in *2001* (the reader will recall they are never shown). Davidson claims, pp. 178-79, that in a 1965 meeting SK told Sagan about his proposal to depict the entities in the film, that Sagan argued against showing them and that it was this advice that 'Kubrick followed.' This makes a good story, but it is simply not true. SK was exploring ideas for the aliens later in 1965 and through 1966 when Christiane Kubrick was still sketching designs and I continued researching Giacometti sculptures (SK was much taken with them), Max Ernst paintings and fantastic art generally, looking for alien ideas. None of this would have been done had SK followed Sagan's advice. SK realised himself in the end that showing them was a Bad Idea, just as the end pie fight in *Dr. Strangelove* was also a Bad Idea and was similarly abandoned. Creative works are judged by what they include, not by what they exclude. SK would consider many things. The 1965 meeting was never to be repeated. SK came up with an excuse to avoid a scheduled meeting for the following day. Davidson suggests, p. 179, that it was due to Sagan's brashness. It was not. SK could navigate brashness with ease when he had to. It was Sagan's patronising manner that did it.

> only more and more so, for four more decades. But no year ever fulfils itself: the cow path of History is littered with the corpses of years, their silly throats slit from ear to ear by the improbable.[10]

SK said he could not have expressed it better himself.

ROGER A. CARAS

Roger A. Caras who conducted the interviews was a thoughtful and generous man with an engaging and affable manner. He was the only other person on the *2001* crew aside from SK who had actually read *À la recherche du temps perdu* from cover to cover, or so he claimed. After serving in the Korean War Caras had a number of jobs in the film industry, including that of Joan Crawford's press agent. He was with SK from the very beginning of *2001*'s pre-production in a variety of roles and later played a major part in the film's advertising and publicity.

He left the film industry in the late 1960s and devoted himself full time to his primary interest: natural history. Caras was widely known on NBC and ABC television reporting on animal and ecology issues and he became president of the American Society for the Prevention of Cruelty to Animals. He was the author and editor of many books.[11] He died in Baltimore in 2001 at the age of 72 (he was born in 1928).

SK'S KEY TEXTS

SK was a voracious and catholic reader from an early age. His interest in science fiction dated back to the 1930s when as a child he read pulp magazines such as *Amazing Stories* and *Astounding Stories* and similar which were then for the ultimate origins of his wish to make a 'really good' science fiction film.

---

10 Bernard Wolfe (1915-85), 'Author's Notes and Warnings,' in *Limbo '90*. The novel was first published in 1953. A masterpiece of dystopian fiction. Whatever became of Wolfe?

11 *Dangerous to Man: Wild Animals – A Definitive Study of Their Reputed Dangers to Man* (Philadelphia: Chilton Books, 1964) has not been superseded as a study of the real and imaginary dangers posed to man by wild animals. A very sympathetic book. Caras also wrote an autobiography, *A World Full of Animals: The Roger Caras Story* (San Francisco: Chronicle Books, 1994).

plentiful on the newsstands of the Bronx. It is to this time that we must look However, he never really considered himself a science fiction fan or buff. It was a genre that he was impatient with: he liked the ideas but he found the writing jejune and the characterisation deplorable. He was the first to admit that science fiction was hardly a subset of *belles-lettres*, as it were, but there was such a thing as *standards*.

It would take a John Livingston Lowes[12] and a lifetime of diligent study to track down the sources and influences that resulted in SK making *2001* but what I would like to do here is list the volumes that SK read and discussed when I first went to work for him in September 1965.

SK thought the two 'best' volumes on ETI were Carl Sagan and I. S. Shklovskii's *Intelligent Life in the Universe* (1966)[13] and Roger A. MacGowan and Frederick I. Ordway III's *Intelligence in the Universe* (1966). SK appreciated having Fred Ordway on hand as the in-house ETI expert to elucidate and elaborate points raised in the book (and much else besides). Both volumes have dated hardly at all and can still be highly recommended.

A good 'once over,' in SK's term, was the book by the *New York Times*' science editor, Walter Sullivan. This was *We Are Not Alone: The Search for Intelligent Life on Other Worlds* (1964, and subsequent editions) and again a volume that still stands up remarkably well.

SK liked the philosophic dimensions and the grandeur of the universe as written about in Harlow Shapley's *The View from a Distant Star: Man's Future*

---

12 John Livingston Lowes, *The Road to Xanadu: A Study in the Ways of the Imagination*, (Boston: Houghton Mifflin, 1927). Over 600 pages of literary detection on the sources and origins of Samuel Taylor Coleridge's *Kubla Khan* and *The Rime of the Ancient Mariner*. This, by the way, was a book that SK frequently dipped into. He admired not just the scholarship but the warmth of the scholarship and the cadences of Lowes' prose style and felt that a book such as this was 'worth a thousand or more volumes of literary criticism with the words "gender" or "exegesis" in the title. The "bull crit" stuff.' Interestingly, SK also liked it because of the footnotes and once showed me what Lowes wrote in his preface, p. xii, 'There are those who find the notes in a book more interesting than the text. I often do myself.' Footnotes, SK said, were where the 'real action' was, where the references were (and he liked references) and where the writer could abandon the sobriety of the main text and not mince his words.

13 Described by Stephen J. Dick as 'the bible of scientific thought on extraterrestrial life,' *The Biological Universe: The Twentieth-Century Extraterrestrial Life Debate and the Limits of Science* (Cambridge: Cambridge University Press, 1999), p. 289.

in the Universe (1963) and he found very provocative Arthur C. Clarke's *Profiles of the Future* (1962) and I. J. Good's *The Scientist Speculates* (1962); the latter he thought a 'Catherine wheel of a book' with its wild and 'partly baked' ideas (the subtitle) and Good's puckish preface which consists of just one sentence which SK found very amusing: 'The intention of this anthology is to raise more questions than it answers.'

Finally, A. G. W. Cameron's *Interstellar Communication: The Search for Extraterrestrial Life* (1963) was an 'early' anthology of some thirty-two scientific papers, some of which were hard going for us in civvy street, but which nonetheless suggested ideas to SK. This was also the volume that introduced SK to the work of Freeman J. Dyson, Philip Morrison and Frank Drake, three of the scientists who are here interviewed.

(Full details of the above titles are given in the Bibliography.)

SK had an abiding interest in ETI and Artificial Intelligence (AI) right up until his death, reading the latest issues of *Scientific American* and getting advance copies of books by scientists such as Marvin Minsky and Hans Moravec.[14]

The reader may care to ponder the suggestion that HAL, the computer in *2001*, can be seen as the father of David, the android boy, in *A.I* (*Artificial Intelligence*), the film that SK bequeathed to Steven Spielberg…

I would now like to outline six subjects that SK found enduringly fascinating before discussing the editing of the interviews.

THE ORIGIN OF LIFE

SK strongly believed that given the size of the universe and given its age it was inevitable that life would arise (through 'chance chemical reactions')[15] and intelligence evolve and that, indeed, there are probably thousands if not millions of civilisations far in advance of our own. He was aware that some scientists have argued that life here on Earth was just a 'fluke' and that the odds

---

14 Marvin Minsky's *The Society of Mind* (New York: Simon and Schuster, 1986) and Hans Moravec's *Mind Children: The Future of Robot and Human Intelligence* (Cambridge, Mass: Harvard University Press, 1988), both had a great influence on SK's conception of *A.I.*

15 'Given a planet in a stable orbit, not too hot and not too cold, and given a few billion years of chance chemical reactions created by the interaction of a sun's energy on the

against it appearing were of an unimaginable order. Nevertheless, we are here and that is proof enough. SK took comfort from the suggestion by David M. Raup, the eminent paleobiologist, that life on Earth may have originated independently on a number of occasions.[16] And why not? He thought this a more likely scenario than every living thing that has ever lived being descended from a *single* chemical reaction in a prehistoric 'soup' ocean.

THE FERMI PARADOX

The Italian-American physicist Enrico Fermi (1901-54), while lunching with colleagues at Los Alamos in 1950, asked the question, 'If there are extraterrestrials, where are they?' This become known as the Fermi Paradox and it is a question that SK returned to time and again over the years. Where, indeed, are they? Why do we have no evidence of any kind for their existence? This troubled SK as he believed firmly in their existence though he was quick to add that absence of proof should not be seen as proof of absence. He saw answering the paradox as an exercise in logic and came up with a number of reasons that go from the comic to the entirely believable:

— We are the only advanced intelligence that has ever existed in the universe. There never have been any extraterrestrials and never will be.

— The extraterrestrials are already amongst us here on Earth or, at the very least, they are closely observing us but they are so smart we cannot recognise them for what they are. They are biding their time before contacting us.

— Extraterrestrials do exist but don't go in for interstellar travel for whatever reason (religion, laziness, biological).

---

planet's chemicals, it's fairly certain that life in one form or another will eventually emerge.' SK, Playboy interview, September 1968. He was aware that this observation applied to life like us and that there may indeed be non-carbon-based life forms in existence.

The *Playboy* interview, along with virtually every other interview SK ever gave, is included in Gene D. Phillips' *Stanley Kubrick: Interviews* (Jackson: University Press of Mississippi, 2001).

16 David M. Raup, 'ETI without Intelligence,' in Edward Regis Jr (Ed.) *Extraterrestrials: Science and Alien Intelligence* (Cambridge: Cambridge University Press, 1985), pp. 31-42.

— Extraterrestrials exist but interstellar travel is impossible (time and/or biological and/or economic reasons).

— Extraterrestrials exist but just have no interest in contacting us.

— Extraterrestrials exist but there are ethical and/or legal reasons why they cannot contact us.

— Millions of extraterrestrial civilisations have existed but something invariably goes wrong before they start seriously exploring the universe (blow themselves up, pollution, over-population, rise of an inward looking religion, etc).

— Only the most violent and evil extraterrestrial civilisations survive. They have detected us and they are on their way now to vaporise the whole Earth (the Buck Rogers Scenario).

— We have been visited in the distant past and we will be visited in the distant future. But right now, everybody is on holiday.

SK, in the last analysis, thought that the immensity of the universe might be the problem, that the distances were just so great that even if a civilisation attained the speed of light there was just not enough time to contact anyone else. Thus great races would rise and fall over aeons and contact would elude every one of them. We are all destined to be alone for eternity and the question, 'Is there anyone else?' would go unanswered for ever.

BRACEWELL AND VON NEUMANN PROBES

In *2001* the black monolith on the Moon is a 'burglar alarm' that when excavated alerts the higher entities to the technological advancement of Earth's inhabitants. Arthur C. Clarke's short story, 'The Sentinel,' which was the basis of *2001*, was written in 1948, and the 'burglar alarm' idea presages by many years suggestions that would be put forward in the extraterrestrial literature of the late 1950s and early 1960s. I'm thinking particularly of two proposals that fascinated SK and though they are not 'burglar alarms' they deal with how advanced civilisations might go about exploring the galaxy, finding out if they were alone and, eventually, colonising space. These are Bracewell Probes and von Neumann Probes.

Bracewell Probes are named after Ronald Bracewell (1921–) of the University of California who first discussed the idea in 1960.[17] Their purpose is solely to observe and establish contact with other civilisations. Manned exploratory spacecraft are costly, whereas a programmed unmanned vehicle is not. So, many of these probes could be built and sent out into the galaxy at considerably less cost. They would be directed at likely planets and would go into orbit and attempt contact. If no results were forthcoming they could move on to other planets or just cruise through the universe looking for signs of intelligence. SK thought it inevitable that any intelligent civilisation akin to our own would produce Bracewell Probes and that we should be looking for them now in our own skies. In fact, telescopic searches have already been done but, so far, with no results.[18]

John von Neumann (1903-57), the computer scientist and mathematician, proposed a theory that machines will one day be able to duplicate themselves, and this is what the von Neumann Probe does. You need only send out one of them. It arrives on a planet and makes several copies of itself which are then launched to other destinations. Each of these probes in turn makes further copies of itself upon arrival and launches them, and so it goes on, exponentially. The 'touchdown' probe in each case, after making copies of itself, would remain to explore and colonise the host planet and, as David Darling has written, it could continue by 'constructing an artificial life-sustaining environment and then implanting this with synthesized fertile egg cells bearing genomes transcribed from the probe's computer memory.'[19] Thus the universe is seeded cost-effectively.

SK thought that any civilisation that produced these probes was effectively committing suicide. If the machines were that smart they would be smart enough to re-program themselves despite whatever safeguards were built into them. He could see them eventually returning to the planet that produced the very first one and colonising *that.* A good idea, he thought, for an Arnie Schwarzenegger movie but not for an intelligent race.

When we look out into space we see no evidence of von Neumann Probes or of the colonisation of the observable universe (as of 2003, that is!). Frank J. Tipler of the University of Tulane noted this and argued in 1981 that the fact that we see no evidence proves that we are alone in the universe. If there

---

17 'Communications from Superior Galactic Communities,' Nature, issue 4726 (1960). For a fuller discussion see Bracewell's *The Galactic Club: Intelligent Life in Outer Space* (New York: W. W. Norton, 1974), Chapter Eight, 'Interstellar Messengers,' pp. 69-83.

18 David Darling, *The Extraterrestrial Encyclopedia* (New York: Three Rivers Press, 2000), p. 52.

19 Darling, above, p. 448.

were advanced civilisations out there they would have sent out the probes and we would see the results of their work. Nothing to see? Then we are alone. Case closed.[20]

Let us just explore the mathematics of the von Neumann Probe for a moment. Suppose each probe only produced two copies of itself, then the progression would be as follows: 1, 2, 4, 8, 16, 32 probes, and so on. Suppose, also, that each 'generation,' that is, the time it takes a probe to travel through space and make its two copies, was 200 years. Then, after 30 generations, the passage of 6,000 years, there would be about 1.5 billion probes in the universe. This prompted Carl Sagan to point out that in just 2 million years the entire mass of the universe would have been consumed by them. (Then what? They consume each other?)

Sagan along with William I. Newman countered Tipler's argument by saying that no responsible, intelligent civilisation would ever produce such a machine.[21] It would, as SK said, be committing suicide.

Tipler's argument surfaced again in the early 1980s when Senator William Proxmire used it to stop funding for the NASA SETI programme.

UFOS AND FLYING SAUCERS (AND A NOTE ON ROSWELL)

A UFO is an Unidentified Flying Object. It is merely a flying object that has not been identified, that's all. It could be a commercial airliner seen at a particular angle with the sun bouncing off it, it could be a top secret military plane, it could be a weather balloon, it could be some known or unknown natural phenomenon. It could be many things. It isn't necessarily a craft from some extraterrestrial civilisation though, of course, a flying saucer, if flying saucers exist, almost most certainly are. If I labour the difference here it is because there is confusion in a lot of people's minds: they see the two as synonymous.

SK saw a UFO and was much puzzled by it. On 17 May 1964 SK, Christiane

---

20 Tipler's paper, 'Extraterrestrial Intelligent Beings Do Not Exist,' is reprinted in Edward Regis Jr (Ed.), *Extraterrestrials: Science and Alien Intelligence* (Cambridge: Cambridge University Press, 1985), pp. 133-50.

21 Sagan and Newman's paper, 'The Solipsist Approach to Extraterrestrial Intelligence,' is also reprinted in Regis above, pp. 151-161.

Kubrick and Arthur C. Clarke had gone to the terrace of 145 East 84th Street in New York (where SK and his family were then living) after working on the *2001* script. It was 9pm. The night was clear and there was moonlight. In the sky appeared a bright object that moved, remained stationary, and then moved again. It was brighter than the surrounding stars and was visible for some ten minutes. SK checked it through his Questar telescope but no further detail was revealed. SK thought that it might be the Echo satellite but after checking in the *New York Times* found that it was not due over New York until a little after 11pm, two hours later.

Later, SK and Arthur filled out the 'U.S. Air Force Technical Information' questionnaire covering unidentified aerial phenomena that was supplied to them by the USAF and submitted it on 16 June 1964. SK appended a typewritten note:

> There seem to be two mysteries here: (1) the stationary period of the object which lasted about two minutes, and (2) the passage of something looking like Echo, almost two hours earlier than the Echo passage.
>
> When the object became stationary it was definitely overhead and my view looked like this:

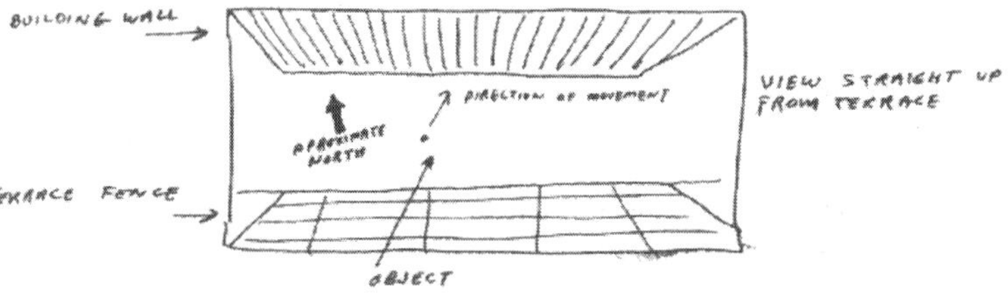

> I considered the possibility I was looking at a star and mistaking it for the object during the stationary period, but the relative speed at which the object had moved prior to the stationary period and after the stationary period would have caused the object to disappear behind the wall of the building within a few seconds. Instead of this I was able to observe the object stationary, using the building and the fence as an excellent frame of reference to indicate the absolute stationary position of the object.
>
> Still looking directly overhead at the object, I was able to observe it begin its movement again, which then caused it to disappear within a few seconds behind the building wall.

> We then carried my Questar 3.5" telescope and tripod, which we had been using on the terrace at the time of the sighting, up to the roof. This took us perhaps a minute and a half to accomplish.
>
> On the roof we were able to find the object very quickly. We set up the telescope on the tripod and managed to locate the object in the lower power viewing lens. With a bit of finesse it was possible to then hold the object in the 80X lens for a few seconds at a time by working the declination controls of the telescope. There were no details to be seen under magnification. It still looked exactly like Echo, just a point of light.
>
> We then watched it until it disappeared on the horizon in a north-easterly direction.
>
> Discussing it afterwards, we assumed it was Echo and simply attributed the stationary period to some unexplainable phenomena. This was obviously not a very satisfying explanation, but believing it was Echo we rejected the stationary observation because it seemed too fantastic and too unexplainable. We then checked the *New York Times* 'Visible Satellite' section and much to our surprise found that Echo was not due to appear until approximately 11.03, about two hours after our sighting.
>
> Naturally this caused a good deal of head-scratching, but we then assumed the *New York Times* must have been mistaken. This assumption was knocked out when we actually observed the Echo at the proper time.
>
> The next day, or perhaps the day after, Mr. Clarke called Dr. Franklyn Branley at the Hayden Planetarium in New York. He told Dr. Branley what had happened and asked him to check the accuracy of the *Times'* information. Dr. Branley called later in the day and confirmed that Echo had passed at 11.03 and 1-something a.m. I do not have a copy of the *New York Times* and I am not sure of the exact Echo passages it printed, but I know it was about 11.00 p.m. and 1.00 a.m.
>
> Although we could find no explanation for this phenomenon we were very hesitant to report it assuming it had some explanation. However, about a week later we mentioned it to Mr. Tom Turner [Republic Aviation Corporation, Washington DC] and he felt the report was definitely worth passing on to the USAF.

What could it be? SK found out that the *New York Times* was in error in reporting the time of Echo's transit over New York as, indeed, was the Hayden Planetarium. The matter was finally resolved when SK wrote to the USAF and got the correct times. It was Echo, no ifs or buts, and the 'stationary period' was no such thing. It was the satellite reaching its zenith and *appearing* to be stationary.

So, the mystery was solved. But, as SK pointed out, had he and Arthur not followed through on this, they would have spent the rest of their lives talking about a UFO at least, and an extraterrestrial spaceship at most, and thus it would have entered the lore of the folk (as opposed to folklore).

In 1968 after the release of *2001* SK had the following to say:[22]

> The most significant analysis of UFOs I've seen recently was written by L. M. Chassin,[23] a French Air Force general, who had been a high-ranking NATO officer. He argues that by any legal rules of evidence, there is now sufficient sighting data amassed from reputable sources – astronomers, pilots, radar operators and the like – to initiate a serious and thorough world-wide investigation of UFO phenomena.

As the years rolled by SK felt that the 'sufficient sighting data' started evaporating until one was left with a residue of sightings that in the future, if not now, would be explained as natural phenomena, secret government projects, or misperceptions. Where was the tangible proof of ET UFOs? Where was the irrefutable film footage? The fragment of metal unknown on Earth? None of this came to pass and instead the anecdotes just piled up and when the abduction stories started circulating SK felt it was time to call it a day. He said the onus was on the believers to prove their claim about flying saucers, not on us to disprove.

Secret government projects? Right. But not hardware, more psy-ops. There was a paragraph in *Report from Iron Mountain* that SK thought had a distinct echo of truth:

> Experiments have been proposed to test the credibility of an out-of-our-world invasion threat; it is possible that a few of the more difficult-to-explain 'flying

---

22 Interview, *Playboy* magazine, September 1968.

23 I think this was from a newspaper article by General Chassin, but I have been unable to locate it. Chassin was a veteran of the Indo-Chine wars. In 1960 Representative Leonard G. Wolf of Iowa entered into the Congressional Record that 'certain dangers are linked with unidentified flying objects.' Wolf quoted Chassin that 'if we persist in refusing to recognise the existence of the UFOs, we will end up, one fine day, by mistaking them for guided missiles of an enemy – and the worst will be upon us.' Chassin was later to write the preface to one of the wackier flying saucer books, Aimé Michels's *Flying Saucers and the Straight-Line Mystery* (New York: Criterion Books, 1958), a work that has had some influence on 'ley' hunters and the New Age 'lost wisdom' crowd.

> saucer' incidents of recent years were in fact early experiments of this kind.[24]

Yes, he thought, the government had been up to some hi-jinks. His reasoning was as follows: we can be fairly sure that after the first saucer 'flaps' in the late 1940s and then onward into the early 1950s the US government was as much in the dark about the saucers as the public. Did flying saucers really exist? Were they from another planet? Were they benign or malevolent? At this stage of uncertainty, SK reasoned, it would be crazy for the government not to look at the 'worst case scenario' and attempt to gauge and research the effects of a possible out-of-our-world invasion on the public. This was basic contingency planning (and so much the better if it also acted as a smokescreen for some of the nefarious things the government was up to).

While we are on the subject of nefarious government activities, let us take a visit to Roswell in New Mexico where, in 1947, a flying saucer is alleged to have crashed. This incident has produced an enormous literature and even films and TV series.

And the controversy shows no signs of abating.

Undoubtedly, something did crash, but what? Was it a flying saucer? Was it some government aircraft or rocket? Was it a high-altitude weather balloon from the secret Project Mogul? Or what?

SK and I developed a theory in conversation that, we felt, explained what

---

24 *Report from Iron Mountain on the Possibility and Desirability of Peace* (Harmondsworth: Penguin Books, 1968) p. 98. 'With introductory material by Leonard C. Lewin.' First published in 1967 by Dial Press in New York. SK became aware that the *Report* was not by a top secret US government group as it purported to be, but was written by Lewin, as was revealed some years later. SK did not feel, however, that this invalidated all of its conclusions.

There is a strong Strangelovian undertow to the *Report*: peace, it argued, 'would almost certainly not be in the best interests of a stable society,' while war was vital for the US and world economy.

Lewin (1916-98) wrote the *Report* after reading an article in the *New York Times* at the time of the Vietnam War which stated that the stock market had dipped after a 'peace scare' (!). SK knew of this and there is an echo in the US Marine colonel standing by the mass grave in *Full Metal Jacket* who says, 'We've gotta keep our heads until this peace craze blows over,' (Stanley Kubrick, Michael Herr, Gustav Hasford, *Full Metal Jacket: The Screenplay* [New York: Alfred A. Knopf, 1987], p. 72).

really went on. It started out as an exercise in 'what would the most off-the-wall paranoid explanation be'? But over the next few days I began to take it more seriously and, I suspect, SK did too though neither one of us admitted it to the other.

Ask yourself this: what could the government be up to that must remain secret for all time, even at the cost of possibly panicking the public by surreptitiously suggesting it was a flying saucer from another world? What had to remain a secret even fifty years later when the Government Accounting Office in 1994 got the USAF to prepare an investigative report (announcing that it was, indeed, weather balloons)?[25]

We would not have thought twice about the 1994 report were it not for one small item that set off alarm bells: the USAF's mention of anthropomorphic 'test dummies.' The reader will recall that several witnesses claimed to have seen the bodies of dead aliens after the crash. Here, for the first time, the government acknowledges that there were bodies, yes, but they were test dummies…

Not far away from Roswell is the White Sands Missile Range where rockets were being tested from 1945 onwards. Could it be that what was going up in them were test *humans*, not dummies? Now, these were infant days in rocketry and the chances of fatalities were high, so we can discount the military seeking volunteers from the services when 'expendable' humans were freely available. And who were these expendable humans? How about 'mental defectives' from state asylums? They would not be required to perform any function in the rocket, they were there merely to have their physiological states recorded by instrument. That might be the secret to be kept at all costs.

If the reader thinks this too far fetched one can point to innumerable cases of state-sponsored abuse of unwitting citizens in the twentieth century.[26]

---

25 Headquarters, United States Air Force, *The Roswell Report: Case Closed* (Washington DC: US Government Printing Office, 1997). The Report was written by Captain James McAndrews. A better argued and more concise demolition of the Roswell saucer story is B. D. Gildenberg's 'A Roswell Requiem' in *Skeptic* magazine (Altadena, CA), vol. 10, no. 1 2003. Gildenberg is a meteorologist who worked on some of the USAAF and USAF projects, including Project Mogul, that form the contemporary backdrop of the case.

26 Start with the US and UK governments' planned exposure of troops to radioactive fall-out for research purposes, take in the 40-year-long Tuskegee syphilis experiments that purposely withheld treatment from Afro-Americans so the course of the disease could be studied, and consider the horrendous experiments conducted by Ewen Cameron at McGill University mental hospital as part of the CIA's MK Ultra programme.

THE DRAKE EQUATION

SK had a lasting interest in mathematics for two reasons. There was the usefulness of *applied* mathematics, in which he had a very good working knowledge, especially of geometry, trigonometry and calculus (valuable for camera lenses, set construction, camera positioning, and so on) and the immutable logic of pure mathematics. He once said to me, 'One plus one is two no matter where you are in the universe… unless, of course, it isn't.'

I remember on *2001* that SK had a large collection of slide rules and was a fast and adept hand at using them. He read somewhere of the great speed and complexity of calculation that could be achieved on the abacus and immediately got a handbook on how to use it, though he never actually got the tool itself ('You don't how good something is until you try it'). Then the first desktop scientific calculators came along in the 1970s, from Hewlett-Packard. One could almost say that SK had been waiting his whole life for them. Now, incredibly complex functions and routines could be performed in the time it took you to hit a few rubber keys. SK never looked back. And PCs were just around the corner.

SK's library was well stocked with mathematical works and books of tables. He was never far away from his copy of *The Van Nostrand Reinhold Encyclopedia of Mathematics*.[27] Another much used volume was Lancelot Hogben's famous tutorial volume on historical principles, *Mathematics for the Million*, the subtitle of which undoubtedly appealed to SK: *A Popular Self Educator*.[28]

I mention all of this to give some background as to why the Drake Equation appealed doubly to SK. First, it was about extraterrestrial intelligence, and secondly, it was an 'elegant' mathematical equation that conferred order and logic on open-roofed speculation.

And what is the Drake Equation? It is a formula for estimating the number of intelligent extraterrestrial civilisations in the universe that have the means to communicate with other civilisations and was devised by Frank Drake, the radio astronomer, for a conference at Green Bank, West Virginia, in 1961 on 'Intelligent Extraterrestrial Life.' Here it is:

$$N = R^* \times fp \times ne \times f1 \times fi \times fc \times L$$

---

27 W. Gellert, H. Küstner, M. Hellwich, H. Kästner (Eds.), New York: Van Nostrand Reinhold, 1977.

28 London: George Allen & Unwin, 1947. First published in 1937.

It is explained as follows:

N = present number of extraterrestrial civilisations able to undertake interstellar communication
R* = mean rate of star formation in the galaxy
fp = fraction of the stars that have orbiting planets
ne = average number of planets suitable for sustaining life
f1 = the fraction of these on which life develops
fi = the fraction of life sustaining planets that evolve intelligence
fc = the fraction that develop the means to communicate
L = the average lifetime of a technologically advanced civilisation

Most of the factors that make up the equation are difficult if not impossible to determine with current knowledge. However, we can be reasonably certain about the mean rate of star formation, the first factor. If we assume that the number of stars in the galaxy is approximately 400 billion and the age of the galaxy is about 10 billion years then this gives us a star formation figure of around 40 stars per year. The next three factors are likely to be known with some certainty in the next generation or two while the remaining three are likely to be speculative for a long time to come. A recent informed 'working out' of the Drake Equation produced a result of fifty civilisations in the entire galaxy actively engaged in trying to communicate.[29]

RADIO ASTRONOMY – SETI

Radio communication is a recent innovation, barely a hundred years old. Who is to say how we will be communicating in a hundred years' time? Two hundred years' time? SK thought the various SETI programmes were a little presumptuous in assuming that advanced extraterrestrial civilisations would still be communicating in this way. Societies at our level of development might be employing radio, but what of those that were more advanced? Is radio the very last word in communication? Is it here for all eternity?

His feeling was that SETI, at best, could only hope to communicate with 'equals,' not peers. He had an analogy with Elizabethan times. News of the

---

29 Darling, above, p. 112.

Spanish Armada's appearance off the coast of Cornwall in July 1588 reached London in about 45 minutes, a distance of some 250 miles. How was this achieved? Simple, beacon fires on hilltops across southern England. Now, supposing the Elizabethans decided to contact extraterrestrials. How would they do it? They would use their most advanced means of communication and thus light a very big beacon fire. SK applauded the optimism of SETI but wondered if it wasn't just that: a mere beacon.

THUS SPAKE STANLEY

For the thirty years after the release of *2001* until his death in 1999 SK and I returned countless times to the subjects spawned by the film. Months might go by with no reference to them and then there would be a newspaper or magazine article or something on television or in a film and we would be discussing them again with current activities temporarily put on the back burner.

I do not wish to imply that these subjects were SK's sole or main concern. They were not. They were just one of the many areas that SK had a keen and abiding interest in.

Others, taken at random, would include Napoleon, Julius Caesar, Hitler, the Holocaust, Ernest Hemingway, physics, contemporary affairs, boxing, American football, the two World Wars, the American Civil War, and so on right down to such arcane areas as the history of interest rates, refinements in the development of the Colt revolver, the lessons of Dien Bien Phu, early influences on Curzio Malaparte, who really discovered America, and English place name etymology, to name but a few.

What I hope I have done here is give some indication of SK's interests (and they were ever widening). He retained right up until the end an intellectual curiosity about the world that most people shed within a couple of years of leaving school (if, indeed, they had it in the first place).

I left school when I was 15 with no qualifications and I went to work for SK a week into my seventeenth year. I was immediately a beneficiary of SK's intellectual adventures. Books and magazines and films would arrive in the office by the wagon load and once SK had gone through them they were left to me. My education had just begun. It was a great education.

• • •

THE TRANSCRIPTS: A NOTE ON THE EDITING

When the 35mm b&w prints of the filmed interviews[30] arrived back from the processing lab, Humphries in north London (a lab no longer with us), they were sent over to the MGM Sound Dept. and audio-cassettes of the soundtrack were run off and given to Vicky Ward, SK's flame-haired PA from Chelsea (she subsequently married the actor Alan Bates), who then began transcribing them on what I think was a Philips dictating system (cassette deck, earphones, foot pedals). Her fingers would be flying over the IBM Executive typewriter and in between she would be fielding telephone calls, running errands for SK, and three dozen other things.

What Vicky was producing was a transcription merely for SK's reference when he would be in the cutting rooms editing the interviews. So, spelling and punctuation were not vitally important. As long as all the words were there in the correct order, that was all that mattered (phonetic spelling of unknown words was OK).

Vicky produced, in the limited time available, a workable transcription and certainly not something that she (or SK) ever thought would be published. Thus, a considerable amount of work was required to get the interviews into publishable form. Vicky was unfamiliar with many words and her punctuation often resulted in an interviewee appearing to say the opposite of what he believed. This is not to cast doubts on her abilities; au contraire. It is just to highlight the gap between cutting room reference and a published book.

I resisted the temptation of heavy editing and reducing everyone to a seamless flowing of silvery-tongued eloquence. These are the interviews as they were given and if, say, Roger Caras or Harlow Shapley sometimes sounds like he's reciting paragraphs from the last words of Dutch Schultz, so be it. I've retained the strangled thought, the aside within an aside, the meandering thought that never achieves resolution. But I have regularised spelling and styling.

Occasionally it has been impossible to figure out what word was spoken and then I've put a question mark in brackets: [?]. When a phrase or sentence has been indecipherable I've put in an ellipsis, three periods: … Caras consistently uses the word mitigate when he means militate. I have changed all those.

---

30 The 35mm film would have been blown up to 70mm by Technicolor in London for *2001*'s initial Cinerama release.

One's mind has a wonderful ability to sort, keep track and order other people's speech. The same words transcribed might appear confused and confusing. So, if you should have any difficulty with the occasional interviews, get someone to read it to you. Your ears will organise it.

All of the following kindly took time off from their busy schedules to further correct my edited version of their interviews and to them I extend my thanks: Jeremy Bernstein, Frederick C. Durrant III, Freeman J. Dyson, Irving John Good, the late Gerald S. Hawkins, Norman S. Lamm, Sir Bernard Lovell, Marvin Minsky, Philip Morrison, and Fred L. Whipple.

The interviews should be read in sequence. I have added footnotes to explain certain references but these expositional notes are not repeated. That is, for example, I explain the term panspermia the first time it appears but not the second.

I have also included at the end an annotated bibliography and a guide to Web resources for the reader who wishes to follow through on the principal matters discussed here.

Jeremy Bernstein, Frederick C. Durrant III, Irving John Good, the late Gerald S. Hawkins, Sir Bernard Lovell, and Philip Morrison all readily contributed their thoughts 'forty years on' for publication here and these contributions follow on immediately after their interview. This had added a further perspective to the book and I am most indebted to them.

[Anthony Frewin was an assistant to Stanley Kubrick on *2001: A Space Odyssey*, 1965-9, and again as an assistant from 1979 until Kubrick's death in 1999. He continues to work for the Kubrick Estate. He has an enduring interest in the extraterrestrial life debate that dates from his time on *2001*.]

Acknowledgements

I AM MUCH INDEBTED to Christiane Kubrick and Jan Harlan who readily responded to the suggestion for this book and made the means available for its realisation. Great thanks must also go to Sir Arthur C. Clarke and Frederick I. Ordway III, both illustrious alumni of *2001*, for agreeing to write the contributions that bookend this work. Further thanks go to Fred Ordway for editorial help and assistance above and beyond the call of duty: what initially was a request for an afterword soon became much more than that. I am deeply appreciative.

I am also greatly obliged to all of the following who were interviewed back in 1966: Jeremy Bernstein, Aspen, Colorado; Freeman J. Dyson, Princeton, New Jersey; I. J. Good, Blacksburg, Viginia; the late Gerald S. Hawkins, Washington DC; Frederick C. Durant, Raleigh, North Carolina; Sir Bernard Lovell, Swettenham, Cheshire; the late Philip Morrison, MIT, Cambridge, Massachusetts; Fred Whipple, Belmont, Massachusetts.

Special thanks also go to Bruce Stiglitz and Louis C. Blau in Los Angeles.

Dr John Braverman SJ at Loyola University in Chicago kindly contributed an appendix that supplies an intriguing 'take' on *2001*.

Father Gene Phillips SJ of Loyola University, Chicago, was, as ever, unfailingly helpful and worldly in his counsel.

Father John LaMantera SJ of the Maryland Province of the Society of Jesus supplied research at a moment's notice as, indeed, did Dr Robert E. Paull at the University of Hawaii, Manoa. I thank them both.

In alphabetical order I would also like to register my thanks to all of the following: Sophie Ball, Reading University, Berkshire; Mrs Jill Caras, Freeland, Maryland; Professor Gerald D. Carr, University of Hawaii; Alison Castle, Paris: Tracy Crawley, St Albans, Hertfordshire; Julia M. Dobson (Mrs Gerald S. Hawkins), Washington DC; Janet Eaton, Jodrell Bank Observatory/University of Manchester; Dr Cleomenes A. Generales MD, La Jolla, California; Andrea B. Goldstein, Harvard University Archives; Andrew Hewson, Notting Hill: Rachel Hunt, Hemel Hempstead, Hertfordshire: Christine Mitchell, Wimbledon, Surrey; Dr Boniface Peiris, University of Kelaniya, Sri Lanka; Robin Ramsay, Hull, Yorkshire; Fabienne Reichenbach, Paris; Diane Richards, SETI Institute, Mountain View, California; Ida D. Schwarz, Yishiva University, New York; Anne Brownell Sloane, The Institute for Intercultural Studies, New York;

Julie S. Virgus, The B. F. Skinner Foundation, New York; David De Wilde, Dorking, Surrey.

Nick Frewin, my son, gave me much computer assistance throughout the months of this book's gestation and Sacha Trusler undertook the onerous job of re-keyboarding the original interviews with speed and accuracy.

The Web generally and the NASA Astrophysics Data System in particular were of inestimable help.

Any residue of inaccuracy, mis-information and nonsense that may litter these pages is entirely of my own devising.

Anthony Frewin

Roger Caras and feline friend – late 1960s?

Isaac Asimov

[Associate Professor of Biochemistry, Boston University School of Medicine, Boston University Medical Center]

SCENE ONE / TAKE ONE

CARAS: Doctor Asimov, do you believe that there is life on planets other than Earth?

ASIMOV: Yes! I believe that there is life on planets other than Earth, but we have to make one thing certain. I do not mean planets in our solar system. The chances that there is life on other planets in our solar system are small. There may possibly be life on Mars but even that would be very simple forms of life. If we are talking about complex forms of life, intelligent life, life that might be roughly comparable to ourselves in intelligence if not in shape, then we are speaking about planets in other stellar systems, planets circling other suns.

CARAS: If a life form arises on another planet and its morphology, its physiognomy is different than our own, but it is intelligent, it can communicate, do we call this life, 'man'?

ASIMOV: It is hard to say how we ought to view life on other planets. So far only one type of intelligent life is known for sure, ourselves. We consider ourselves men. We suspect that perhaps porpoises may be reasonably intelligent but even if it turned out that they were as intelligent as we are we still wouldn't call them men. We might have to invent an entirely new word to generalise different species that have in common high intelligence. What could we call them? I don't know. Could we call them 'tellies' for instance, meaning 'intelligent beings'? But it is not important to call anyone anything. It is the concept we have. I look forward to a time myself, almost certainly not in my own lifetime, to a time when we will recognise not merely a kinship of all men but a kinship of all intelligent beings. It seems to me that all intelligent beings will have a common enemy, which is ignorance. The universe, the hostile universe around us, it would be very nice, I think, if the time would come some day when these 'tellies' would stand together, really find room in the universe for each of them without impinging halfway on each other and fight together against ignorance.

CARAS: Some science fiction writers in creating their stories of life on other planets have made them look a little hostile to life on Earth. Is there any reason on this Earth or in our imagination to assume that any other form of life would necessarily be hostile to us, or even likely to be hostile to us?

ASIMOV: We have no experience really of other intelligences and therefore it is hard to predict whether they would be friendly or unfriendly. In writing science fiction one has a tendency to make these other intelligent creatures unfriendly sometimes because that adds drama to the story. It adds conflict. But I wonder whether this would necessarily be so. If we found another race of intelligent beings I'm sure that we would be most interested in studying them rather than in killing them. They, I hope, would be more interested in studying us than in killing us. If we are both intelligent it is almost certain that we will know something that they don't know, that they will know something that we don't know. It will be almost certain that our views and ways of thinking might have certain interest to them and vice versa. In short, there would be a great many reasons for self-interest in friendship. Simply to go out on a mad career of conquest does not seem reasonable, not at first, anyway. Merely reaching another world would be such a terrific undertaking that we would probably get vastly under strength, so to speak, but supposing one ship gets there, is it going to conquer that world? Even if we come back are we going to send out an armada to conquer it? By the time we can do that in any reasonable fashion many years will have passed during which time it would be much more interesting if we studied them and learned from them and vice versa and by the time we did that we might find that there was plenty of room to expand in without impinging on other intelligences, and they would find out the same thing.

CARAS: Most scientists believe that there must be life and probably intelligent life elsewhere in the universe.

ASIMOV: My belief in the presence of intelligent life elsewhere in the universe is purely a matter of statistics. There are 135 billion stars in our galaxy. Now, that is just in our galaxy. There are at least 100 billion other galaxies like our own, more or less. Among these 135 billion stars a certain moderate percentage are like our own Sun. Of these stars almost all would be expected to have planetary systems. A certain moderate percentage would be bound to include an interplanetary system, a planet very much like our Earth. On such a planet life would be almost sure to develop, because scientists don't think that life is an accident, a miraculous accident. It follows almost inevitably from certain well-known chemical changes. Well then, it is estimated, it has been estimated by somebody, a man called Stephen H. Dole at Rand

Corporation who put a lot of time into working these things out, it has been estimated on the basis of what we know or can reasonably guess that there may be as many as 640 million planets very much like our Earth, sufficiently like it so that they could be inhabitable.[1] We could get there in a spaceship and just walk out without protection. There are 640 million of these in our galaxy alone. Now, on these 640 million, even if we assume that only one in a million is likely to develop life to the point of intelligence, there would still be 640 intelligences, not just life but intelligences in our galaxy. If we assumed that we are nothing special, we are just kind of average intelligent, there would be 320 less intelligent than us and 320 more intelligent than us and, as I say, I repeat, only in our own galaxy alone there are a 100 billion others. So it seems unavoidable on the basis of statistics alone (a) that we are not alone, that there are a lot of life forms, and (b) that a number of these planets with life forms on them have intelligent life on them.

SCENE TWO / TAKE ONE

CARAS: We'll get back to intelligence in a minute, but I'd like to have on record your views as to the origin of life here or on another planet.

ASIMOV: Life originates on Earth, or any planet. Life originates out of the chemicals that exist on such planets. Now, the ordinary chemicals, the ones that are most common, on any developing planet would be hydrogen, helium and the number of compounds of hydrogen with carbon, oxygen, nitrogen. In other words, we think we know just what simple compounds would be present at the beginning. Now, if we take mixtures of these compounds of water, methane, ammonia, hydrogen and mix them under conditions where energy will flow into them, from an electric spark, for instance, then after just a matter of weeks more complicated compounds will build up and these will be similar to those compounds which form the basis of the complicated molecules of life. In other words, we can sort of follow these simple compounds on a path part way toward the complicated molecules of life and we have followed them up to very complicated molecules, one in particular called ATP[2] which is the key compound in living things. In all living things

---

1 Stephen H. Dole, *Habitable Planets for Man* (New York: Blaisdell, 1964).

2 Adenosine Triphosphate. The 'key compound' in all living things. It is the molecule that transforms energy into a form that the organism can handle.

we know of as far as the use of energy is concerned, nowhere have we discovered anywhere where this path veers away from our direction. In making these things happen in a test tube we seem to be on a road which will end in life as we know it. This makes us think that this is the road actually taken at the beginning. They didn't have just a few quarts of material to work with. They had a whole ocean of chemicals to work with. They had a billion years to work. So, we are convinced that they are just a matter of the random motions and combinations of atoms that slowly move and more complicated molecules were built up until finally we had some that were complicated enough to be able to reproduce themselves and once we had a self-replicating molecule as we can call it, then we had the fundamental basis of life, all else is just variations on a theme, just commentary: and this, according to our present thinking, is bound to happen on any planet which has a chemistry similar to that of the Earth.

CARAS: Can you envision life coming into being on any planet and being static, not going through an evolutionary cycle once it was there?

ASIMOV: Once life has formed it must evolve because none of the processes whereby molecules replicate themselves, produce others like themselves, is, as far as we know, perfect. Nor can any such process which involves just random physical and chemical behaviour, nor can it be perfect. There is always the chance of little mistakes in the process of what an information man might call noise and so long as that happens this means that even if you start off with a group of molecules all identical eventually you will have a spectrum of molecules which will vary amongst themselves. Some will be 'better' than others in the sense that they will perform whatever function they are performing more efficiently. These will drown out the others to a certain extent. Now, it may be that on one part of the Earth one type of molecule will be more efficient and on another part, with a different environment, different temperature and so on, another type will be more efficient. In this way you will eventually develop different types of molecules here and there and perhaps as these molecules develop into something still more complicated, like cells, the same process will take place and you will have one kind of cell here, one kind of cell there. In short, evolution is bound to take place anywhere that life develops.

CARAS: When an evolutionary cycle begins, given enough time, is intelligence the inevitable result?

ASIMOV: We can't say that intelligence must result whenever we have life. Here on our own planet, as far as we know, life may have continued for something like as much as three billion years with nothing that we would recognise as intelligence until perhaps a million and a half years ago, a length of time which is almost unimportant compared to the total length of time that life has existed. Now, if we could somehow imagine ourselves to be a disembodied spirit looking down on Earth as it was say, two million years ago, we would say here life continued for three billion years, no high intelligence has developed, therefore we might very well reason that there is no such thing as high intelligence, you see it will never develop so this path which took three billion years perhaps to reach intelligence here on Earth might, on another planet, take five billion years for it to develop if it takes a different turning slightly and therefore on that planet if life has only existed three billion years there would be no intelligence yet. On the other hand, some other planet might be lucky and develop intelligent life after only one billion years and then it might have several billion years to develop further to achieve heights we can't imagine, perhaps to destroy itself. So we might guess that there is a chance that intelligent life may exist elsewhere but we have no way on the basis of the knowledge we now have, we have no way of predicting how likely that chance is, whether it would happen one out of every ten times, one out of every hundred times, one out of every million times.

CARAS: Let's assume for a moment there is a planet where life is naturally more advanced than ourselves or that it has developed intelligence and that instead of having roughly ten thousand years, it has a civilisation, say of four million years. What might some of their capabilities be?

ASIMOV: Gee! When one speculates about what an intelligent race of beings could do in four million years one has to think with care. I would like to suppose that what we have not yet done is to work out relatively completely the mechanisms of the brain. Then we will finally know just how their mind works. That would mean that they would understand themselves completely and be able to wipe out mental disease. In other words, they would be a race of mentally healthy beings. Also, that they would be able to construct devices that would duplicate the workings of the mind and perhaps even improve on it in some way. So, in other words, it would be a race that would be well served mechanically by what is called robots that would do all the mental scuff-work so to speak, leaving for the intelligence themselves the various very highest forms of intellectual labour that is creativity of one sort or another. So I would expect that a race that existed for four million years, intelligent all the time, might not be high in numbers, they wouldn't have to be, but they would be high in quality.

CARAS: Would you speculate on what might some of our capabilities be in the next ten thousand years?

ASIMOV: Ah, well, in the next ten thousand years we will have the solar system completely explored. We will have colonies on the Moon or on Mars, maybe on the moons of Jupiter. I hope that we will also be living on the continental shelf. I hope that we will have controlled population, that by that time we will have reached some level of population which will be suitable for the world as we want it to be and we will keep to it and I hope also that we will reach the point where we can treat mental illness at least as well as we can now treat physical illness. I would like to see a happy, well adjusted world.

Well, now, the thing that bothers many scientists now is the science itself in the sense that results are being accumulated, new knowledge is being established so rapidly, that it's becoming more and more nearly impossible to keep up and something brand new must be devised to keep on top of their jobs. What seems most logical now is, for the moment at least, the use of computerised systems of information retrieval. Only today I saw my first paper which had as part of its abstract, the first scientific paper which had keynotes and then a series of key words from the paper itself, these key words are used as places under which to file the paper in some sort of computerised memory system. Then if you want to know all the papers that have as one of its key words intestinal cells or metabolism or something like that, you press the proper buttons, so to speak, and the whole pile of papers comes out. This is a kind of mechanised automated information retrieval that we need, but it won't be the last word either. What do we do beyond that? I like to think, if I may just wander into the wide blue yonder, that the greatest computer we are ever likely to have is the human brain itself. If, in order to make a mechanical computer or even an electronic computer as efficient as the human brain and as compact, we would have to work with units that are themselves as complicated as the individual brain cells and use about ten billion of them. Well, if we are going to do that, why not make use of the computers we do have with units that many, that small, that complex. In other words, our own brains.

I look forward to a far future perhaps in which it will become possible to hook brains together in tandem so to speak, somehow, where we don't have things tackled by a single brain at a time but perhaps a team of large numbers of brains in parallel. Whatever any one of those brains knows or remembers can be tapped by all the brains. This is not completely a pleasant thought to me since I value my privacy, but you know things can change, people can get used to it, maybe even the next step in evolution, the multi-human being, so to speak, and maybe if we meet very advanced civilisations on other planets this might be something that we'll find. No such thing as individuals in our sense. We've got a race being, you know – One for All and All for One.

CARAS: Do you think we will one day have machines, aside from the tandem people, that are as intelligent as human beings?

ASIMOV: Yes, but it is very difficult to say whether machines will be as intelligent as people without qualifying. Is it necessary that machines be as intelligent as people in every possible way? Do we really want to duplicate a human being completely, mechanically? Supposing we have a machine that is smarter than a human being in one particular way, another machine that is smarter than a human being in another particular way? It might be more economical to specialise our machinery in ways that human beings don't specialise. Keep human beings for their particular speciality, which is lack of specialisation. In other words, the human brain is very good in many different directions and it would pay us to supply individual machines for individual capabilities just as we do in our machines now which mimic man's physical abilities. We have the capacity for doing a great many different things. But if we have a machine which just cuts an object as it goes past, it cuts the object more exactly and more rapidly than we could possibly do by hand but that is all the machine can do. We can do a trillion other things you see and it may be the same exactly with machines designed to mimic our mental capacities.

CARAS: You talked before about tandem people. If you link two human beings together, their intelligence and their personality, they are no longer individual people. What is it – a man? Or is it a biological machine?

SCENE THREE / TAKE ONE

CARAS: We're facing, very quickly now, an inundation, an intelligence explosion. Is the super-intelligent machine which we are told will be here, is it going to be an electronic or a biological machine?

ASIMOV: Well, it is neither a man nor a biological machine. Perhaps it is the next step, the next important stage in evolution? We have started perhaps in an evolution with individual molecules equivalent to very simple viruses. We have perhaps graduated from that to cells which are complexes of large molecules all co-operating, the individual molecules of which can no longer exist separately and then the second great stage was when cells got together to form multi-cellular organisms such as us. We are composed of something like fifty trillion cells. The fifty trillion cells are no longer truly individual cells. If they were separated they cannot live by themselves. Now we are no longer a cell. We are a man, but we are not sorry about it. We do not regret that our individual cells are no longer individual amoebas or something because we feel, very rightly, that the combined structure of all the cells is something higher and better than those same number of cells separately. And now for something like 600 million years at least we have simply been reading variations on this multi-cellular theme, but we have reached the point now it seems to me where the multi-cellular being has gotten essentially as far as he can go. If he wants to advance really farther there has got to be another great change and maybe the multi-organism creature, the tandem human being is the answer, though we couldn't call it a man, perhaps. We'd have to have some other word. And how such a multi-organism being reproduces itself, maybe it is immortal, you know? Maybe the individuals consisting, making up the tandem die and are replaced but there is no lapse in consciousness? Just as if the cells in the outer layers of our skin gradually die and slough off and new ones replace it. But we are not aware of that. We still have the same skin we always had you see, in the same way the tandem beings may be, as tandem beings, immortal.

CARAS: Do you believe that one day man will have biological immortality? This way, if not another?

ASIMOV: I think this way is the best way, because you can reach a kind of stability. In any other kind of immortality you will practically forbid reproducing of cells because what are you going to do with the new creatures? Here, you have immortality of the race as a whole not of the individuals making it up. In a sense we have it now but there we would have it with a continuity of consciousness which is different.

CARAS: One last word. Cultural shock. How are people going to react if we make contact with extraterrestrials?

ASIMOV: Well, fortunately we are being prepared for it. It won't happen for quite a while, I'm convinced, and through science fiction, even through UFO sightings, the general consciousness of mankind is being prepared for such things and I think that is a very good thing. Maybe even good science fiction movies will help, too.

Jeremy Bernstein

[Associate Professor of Physics, New York University]

REEL ONE / TAKE ONE

CARAS: Jerry, do you personally believe that there is extraterrestrial life?

BERNSTEIN: I think that the scientific probabilities are exceedingly high that there is extraterrestrial life. I think that it would be very foolish for a scientist to make categorical statements either about the existence or the non-existence of extraterrestrial life but, examining the evidence, I would say the probabilities are exceedingly high that there is.

CARAS: Is there anything we know, are there any scientists who know something that militates against its existence?

BERNSTEIN: I think that it is impossible to make any coherent argument against extraterrestrial life as certainly there is nothing, no scientific information that makes extraterrestrial life contradictory, unlikely or impossible from the scientific point of view.

CARAS: What scientific knowledge are we in possession of that would seem to indicate that there is extraterrestrial life?

BERNSTEIN: I think the scientific argument for extraterrestrial life has to be divided into parts. In the first place the universe is so enormous that there are without any question billions of planetary systems and among these planetary systems there are certainly very likely to be some which are capable of supporting life so that there are many, many theatres available on which the drama of life could be played out and the really difficult question is not so much the availability or the probable availability of environments for life, but to try to make some kind of coherent argument, which is more the job of a biologist than of a physicist. As to how life originates from inanimate matter, what the biologists have succeeded in doing in the twentieth century as far as I can see is to narrow the gap between living and non-living material to such an extent that the distinction between living and non-living material has almost become a meaningless question. So, I think the biologists would agree that life can originate spontaneously from non-living material and so that this is presumably possible elsewhere as well.

CARAS: What are your ideas about the origin of life?

BERNSTEIN: I have one statement I would like to make.

SCENE TWO / TAKE ONE

CARAS: Before we go into the origin of life…

BERNSTEIN: There is certainly a great deal of poignancy attached to the search for life elsewhere in the universe: poignancy which is undoubtedly conditioned by our own experience with death. We value life here on our planet. We treasure it because we are all subject to death and one of the most impressive which a quest into extraterrestrial life – if we ever discover intelligent civilisations and communicate with them – is to try to learn from them how they have dealt with death. What forms death has for them. One of the things that have struck me enormously in thinking about life on the Earth is that our life spans are measured in terms of years. It means that if you take the average biological structure on the Earth its life span is about one year, but one year is the time it takes for the Earth to go around the Sun. That is to say it is four seasons, so one can say that the average life span on the Earth is about four seasons. There is something very striking about this to me. I don't know why this should be, but one can imagine that if there are other planetary systems elsewhere whose years are very much longer than our own, or very much shorter than our own, that the intrinsic life span could be very much longer or very much shorter, and so the whole attitude and the whole approach to death could be completely different for such an extraterrestrial civilisation and I myself would think that that would be one of the most striking things that we could learn.

CARAS: What are your views as to the origin of life on Earth? Or any other planet?

BERNSTEIN: I think that the details of the origin of life which is primarily a question of biology, the specific details of how life has originated, must be extraordinarily complicated and may well be something which we will never be able to work out step by step in a scientific fashion. The most important thing to understand in thinking about the origins of life on Earth is that the environment of the Earth when life originated was entirely different from what it is now. In other words, if you try to imagine life originating in our present environment which has a lot of oxygen in the atmosphere which is very chemically reactive and things like that, one is certainly misled and as far as I know it was Darwin who first understood that to think about the origins of life on Earth you have to try to imagine what the Earth was like before life originated and so while I don't know the specific mechanisms by which life emerged from non-living matter, I think it is important to understand in that argument this point about the Earth being very different at that time.

CARAS: Is there anything about the Earth at that time that made it a unique environment in the universe or is our picture of the universe at that time pretty much what we could expect in some of these billions of other galaxies and the billions of other planetary systems within these galaxies?

BERNSTEIN: The entire argument which attempts to generalise from our history of the solar system to what goes on in other planetary systems is really based on a probability that as one first attempts to establish that the evolution of our solar system is not in any way unique. This is, of course, from an exact point of view, impossible to do. We simply do not know enough to state whether our solar system is or is not unique.

What we do know is that there must be billions of other planetary systems. We know that there are stars that have planets. We even know that some of those planets are comparable in size to our own Jupiter and therefore we are sure that there must be billions of planets in the universe but all the rest of it is a conjecture which is based on probability, on the fact that the evolution of life near the Earth, on the Earth, appears to have been a rational, explicable process.

SCENE TWO / TAKE TWO

CARAS: Do you want to make a compact statement about the New York University?

BERNSTEIN: Yes, I will try. I recently gave a physics colloquium at New York University on the subject of extraterrestrial life and before I started the talk I asked the audience whether they believed that there was extraterrestrial life and much to my surprise every member of the audience raised his hand in agreement that there was extraterrestrial life and I think if I had been in the audience I would also have raised my hand in agreement but whether we will contact them and whether that will be soon, it would, of course, be impossible for me to say.

CARAS: What do you think the reaction of the human population on Earth would be to contact with extraterrestrials?

BERNSTEIN: I think the reaction to contact with extraterrestrials would depend very much on how the contact was made. I feel sure that the first contact, which will probably be electronic, will be a very spooky affair and it will be like encountering a certain kind of undecipherable hieroglyphic I expect. In other words, what I think will happen when there is a first contact, if there is a first contact, is that we will see some kind of symmetrical pattern in a bit of electronic communication and it will turn into a kind of modern Egyptology to decipher what it means. I think it will be a very spooky experience for many people and it may just shake us up out of our present tendency for political self-destruction. On the other hand human beings are very terribly adaptable and they may adapt themselves to that idea also very rapidly and go on as if nothing had happened.

CARAS: Is there a reason to believe that some civilisations in the universe might be very much older than our own? What might the capabilities be of a civilisation with four million and ten thousand years of script and mathematics instead of ten thousand years?

BERNSTEIN: It is very hard to predict the future of science and technology. As someone remarked, 'If anyone knew what future science was going to be it would be present science.' But certainly rapid space travel will be quite common unless, of course, there is a fundamental paradox in intelligent beings. What has struck me is that the real future of space travel lies in making use of nuclear explosive propulsion systems. Making use of the energy from controlled nuclear bombs. Therefore it has seemed to me that the time in which a civilisation will learn to do really significant space travel and the time at which they will learn about atomic and hydrogen bombs would probably be about the same time. Then the question is whether it is possible to exist in a civilisation which is sophisticated enough to survive its own technology? We don't really know whether we are going to survive our own technology and at least it's one gloomy possibility that every civilisation which has come up to that technological cleft stick has not been able to survive its own technology.

CARAS: What is the future of the super-intelligent machine? What's it going to be like? Is it going to be an electronic or is it going to be a bios?[1] When are we going to have it and when is it going to be as intelligent as human beings and when is it going to be more intelligent? In three words, please.

BERNSTEIN: In three words! Well, in three words, 'I don't know'. What is, of course, impressive about the human brain as opposed to a computer is the incredible compactness of the brain. The brain involves billions of functioning elements whereas the ordinary computer has at best only millions of functioning elements and the organisation of the brain and the organisation of the computer are really at the moment quite different. The brain

---

1 Caras uses the term 'bios' throughout the interviews to mean an organic computer or machine. It may be a neologism of his own devising.

is set up to process lots of information in parallel. The computer works serially on this information and I think it will be quite a while before a computer is like a brain or in fact the brain like a computer.

CARAS: Will it be an electronic device or will it be a biological one?

BERNSTEIN: Well, I don't know how to answer that. What is striking about the brain is the apparent random element in its organisation. The fact that at least part of the brain's wiring appears to be done according to an essentially random pattern and it may be that the next step in the organisation in the machines will be when one learns how to also construct them in a random or semi-random way.

CARAS: Will the time ever come when there will be a cloudy margin between man and machine?

BERNSTEIN: My guess is yes. I think the question of whether there will be a blurred distinction between man and machines is in some ways rather similar to the question of the distinction between living and nonliving material. The lesson of modern biology is that the distinction between living and non-living material is almost arbitrary. And so it is possible that one would be able to make machines biologically out of test tubes rather than in an electronics factory and then it will be almost an arbitrary question as to whether one wants to call such objects machines or living animals.

CARAS: One day, if we encounter extraterrestrial life which has intelligence equal to or superior to our own, what if the morphology is totally different? What if the physiology is totally different? What if we get a rattlesnake with a turtle shell and an elephant's head but it is as intelligent as we are? What do we call it? Do we call it man?

BERNSTEIN: It may happen if and when we encounter extraterrestrials and we will come upon a curious version of our race problem in a sense that we will come upon species which look terribly ugly, inferior, distorted, from our point of view which, in fact, will have remarkably superior intellectual constitutions and then we may have a problem in reconciling ourselves to such species similar to the kind of problems that human beings have in adjusting to differences in human life, which certainly may be minuscule compared to the kind of differences we will encounter between human life and extraterrestrial life.

SCENE FOUR/ TAKE ONE

CARAS: Do you think male extraterrestrials may… What do you think they will look like, Jeremy?

BERNSTEIN: Well, there's a song about elementary particles which has a punch line, 'A very small rest mass and no charge at all.'

CARAS: Professor Hawkins did a very careful analysis of Stonehenge and found that the machine is an extremely accurate observatory as well as a computer. This has strengthened the argument for many that the planet Earth has been visited in the past by extraterrestrials. I am not talking about UFOs. What do you think? Is there any chance that Earth has been visited by extraterrestrials?

BERNSTEIN: As far as I know there is absolutely no evidence that the Earth has been visited in the past by extraterrestrials. I think one of the most interesting aspects of the exploration of the Moon and later on of Mars will be its archaeology, if any, because of the low atmosphere. If anybody had visited the Moon or had left Mars at an earlier time their remains should be extremely well preserved since there is nothing much in this atmosphere of Mars to have corroded them, but I say there is really no evidence at all that the Earth was visited at some prior time by extraterrestrials, although hope springs eternal.

CARAS: History?

BERNSTEIN: I have been very interested in the history of the subject of the quest for extraterrestrial life. As the first reference as far as I can see to the possibility of extraterrestrial life, at least in the Western civilisation, goes back to Romans Plutarch and Lucian who speculated that there might be life on the Moon. The Romans also thought that the comets might be the souls of dead people of great importance but so long as Aristotelian cosmology dominated the thinking of the scientific world there was a sharper distinction between the laws of nature that applied to the Earth and the laws of nature that applied to the rest of the cosmos, therefore if you have such a distinction clearly the quest for extraterrestrial life is essentially absurd. It was Galileo and the Renaissance scientists in general who broke down the distinction between terrestrial and extraterrestrial laws of nature and therefore it is not surprising that somebody for example, like Kepler, would be very concerned with life in particular on the Moon. Kepler wrote a fascinating book called *The Dream* in which he discussed life on the Moon.[2] The

---

2 Edward Rosen, *Kepler's Somnium: The Dream, or Posthumous Works on Lunar Astronomy* (1634), (Madison: University of Wisconsin Press, 1967). This is a new translation by Rosen who also provides a full commentary. The best edition.

modern thinking about extraterrestrial life has of course taken advantage of Darwinian evolution. It was assumed up until the experiments of Pasteur that life could evolve spontaneously from non-living matter although it was usually assumed at the same time that man had divine origin. Well, of course, if man had divine origin then the question of extraterrestrial life can no longer be discussed in a scientific way. It becomes rather an arbitrary one. But Pasteur's experiments appeared to him to prove that life could not be generated from non-living matter which he took to be a triumph for science. Well, if life cannot be generated from non-living matter then life can only generate itself and so you are led to an infinite regress where all hope of scientific explanation of the origin of life becomes impossible, so it is only with Darwin and his successors that the problem of generating life from non-living matter has been re-examined and therefore the thought that there might be extraterrestrial life is in a certain sense a rather modern thought and rests very much on this blurring of distinctions between living and non-living matter.

CARAS: Is there any logical, coherent scientific argument that would seem to indicate that there is no life other than on this planet?

BERNSTEIN: I cannot imagine at the present state of science a convincing argument which could prove in any sense that there could not be extraterrestrial life.

CARAS: What else would you like to add?

BERNSTEIN: Well, I don't know. I think I have covered everything. Did Dyson talk about his chilling view that extraterrestrial life may not be a benign thing?[3]

CARAS: No, but that is a good point. Let us take that very quickly. Science fiction writers have for a long time projected extraterrestrial life, very often displaying it as something [?]. Is this not a projection of our own insecurities? Is there any reason on this Earth for us to believe that extraterrestrial life would be in any way hostile to us?

BERNSTEIN: I think it is generally and rather naively assumed that extraterrestrial life will be benign. That there will be philosopher kings reigning in the cosmos. I think it was first Professor Freeman Dyson who made the point that if we detect signs of extraterrestrial life what we'll detect first will really be advanced technologies. We will detect signals generated by some very

---

3 Freeman Dyson. Interview, *infra*.

advanced technological system and it is by no means clear that a civilisation capable of advancing in technology is going to be a very benign civilisation. The more technological our own civilisation becomes in many ways the worse it becomes, and one could imagine terrible civilisations with a very high degree of technology.

CARAS: Don't you think that a civilisation that had a few extra hundred thousand or even million years over us could have cleansed itself by then? Put itself on the couch and analysed itself?

BERNSTEIN: Well, one might hope that a civilisation which has been going for hundreds of thousands of years might have worked out all of its aggressions and internal problems but I think it is just as likely to have blown itself off the face of the Earth. I don't know. That's one of the fascinating things that either we or our protégés will learn.

CARAS: The super-intelligent machine when it arrives, will it have a personality and therefore will it have personality disorders?

BERNSTEIN: I don't really know, Roger.

ADDENDUM: Jeremy Bernstein, 2003

In reading this nearly forty years later several things strike me. I was, at the time, at NYU but left soon thereafter. Part of the reason was that writing *New Yorker* profiles of people like Stanley was thought to lack gravitas.

I have no recollection of giving a colloquium at NYU on extraterrestrial life but I guess I must have. I do recollect the occasion of the interview. A British film crew set up in my apartment and managed to blow out all the fuses. This did not sit well with the building management. I had forgotten that Roger did the interviewing, but who else could it have been? When I arrived in London after the interview Stanley insisted that I watch myself on film. I was horrified. I twitched and blinked in an appalling fashion. I looked somewhat deranged. If the film ever resurfaces please don't show it to me!

Frank D. Drake

[Center for Radiophysics and Space Research, Cornell University; Director Arecibo Ionospheric Observatory]

ROLL ONE / TAKE ONE

CARAS: Doctor, do you believe that there is intelligent life in the universe, other than on our planet?

DRAKE: I think it is virtually certain that there is intelligent life elsewhere in the universe.

CARAS: What reason do you have to believe this?

DRAKE: The reason why one believes there is intelligent life elsewhere is based on a body of knowledge which we have accumulated. This includes, particularly, the vast number of stars that exist in the universe: some 100 million, million stars, the fact that most of these are like our Sun, the fact that we believe the existence of planetary systems is a very common thing in the universe and, lastly, that the development of life on planets is not a difficult thing but rather something that occurs very easily when conditions are appropriate.

CARAS: What might some of the capabilities be of an advanced civilisation? Say, one that was a million years older than our own?

DRAKE: I think we are not capable of predicting what the capabilities would be of a civilisation even a thousand years ahead of our own, perhaps not even a hundred years ahead of our own.

CARAS: Does life, when it occurs, where it occurs – does it necessarily then go through an evolutionary cycle? Can life be static and in that same context if life dces go through an evolutionary cycle, is intelligence inevitably the result of an evolutionary cycle?

DRAKE: I think it's clear that life will go through an evolutionary cycle every where. This is forced upon life because no matter what planet life appears upon there is going to be a limited supply of sunlight and therefore a limited supply of food and this inevitably leads to a competition between species for this limited food supply – leading to the development of species which are capable of competing better for the available food. Now, it is more controversial as to whether intelligence is an inevitable result of this evolution.

Some people believe it may not always appear in the course of evolution. However, if one examines the fossil record – the history of life on earth – one sees that only one characteristic has continuously developed and improved throughout the history of life and that is intelligence. Sure enough, animals have tried different numbers of legs – we've had six and eight legs and a hundred-legged things – we've had enormous creatures such as the dinosaurs and little ones. We've had winged creatures – everything has been tried. The only thing which has persisted and continuously developed is intelligence and this argues strongly that intelligence will appear everywhere where life evolves.

CARAS: Does there exist a single reasonable argument against the existence else where than on Earth of life and even intelligence?

DRAKE: There are no reasonable arguments that would lead us to believe that we are the only abode of life in the universe.

CARAS: What do you suggest is the origin of life on Earth?

DRAKE: Well, we seem to have really quite a good feeling as to how life appeared on the planet Earth. It seemed to be an inevitable consequence of very simple chemistry which took place in the earliest stages of the development of Earth. Most people don't realise that the early Earth was much different to what we have now. The atmosphere consisted of quite different gasses, we had the oceans and we had the sunlight. Given just these things alone one can easily show in the laboratory that the basic chemicals from which living things are made would have been formed in great quantities on the primitive Earth and would have evolved, just as living things evolve, into more complicated forms until finally one thing appears which we would call alive, which would have the characteristics of life.

CARAS: Some people talk of the possibility that life in some spore form is to be found in cosmic dust and that life may not in fact have originated on Earth but may have come here in spore form from elsewhere and that life is in fact of extraterrestrial origin. Is there any basis for believing this?

DRAKE: One cannot rule out the possibility that life on Earth in fact was carried to Earth from elsewhere on a small meteorite or as some kind of a spore. In fact there is a theory that this is how life came to Earth called the pansper-

---

1 Panspermia is the theory that life arrives 'ready-made' on a planet from outer space as 'spores,' either blown by cosmic winds or borne on a meteorite or some other body. The idea has classical roots but was first seriously proposed in modern times by the Swedish

mia theory.[1] However, it is very unlikely that this happened because the environment of space is so rigorous. Anything travelling across space is bombarded with cosmic rays, ultraviolet radiation, in such intensity that anything, any life as we know it, would be destroyed over the aeons of time which are required for the transport across space. For these reasons we think it unlikely that life was carried across space from one planet to another.

CARAS: If there is intelligence in the universe and there presumably would be intelligence more ancient than our own, there would be civilisations capable of technological feats very much ahead of our own. Do you suppose that any of them have ever visited us? I'm not necessarily talking about recently, I mean in the past?

DRAKE: That's a very interesting question as to whether Earth has ever been visited by another intelligent species. We can compute that the chances of this happening are very small and that making some rather poor estimates one can guess that this might happen at best once every ten thousand years or so, so that we should not expect frequent visits from other planets. By the same token we cannot rule them out. Now, if they happened thousands of years ago and all evidence of it has been destroyed, one thing does seem certain and that is that we have never found it yet on Earth, any clear-cut evidence of an extraterrestrial visit.

CARAS: Does our atmosphere and all the muck that floats around this planet in it – the debris that we put up – make our planet very, very difficult to observe visually from elsewhere?

DRAKE: Our planet is extremely easy to observe from elsewhere. It is in fact

---

chemist, Svante Arrhenius (1859-1927), in 1903. Many scientists felt that stellar ultraviolet would kill any such organisms. The theory was largely discarded but has recently been revived concerning meteorite ALH84001. This is the meteorite found in the Antarctic that some 13,000 years ago was blasted off the face of Mars by the impact of an asteroid. Some scientists believe that microscopic 'worms' within the meteorite are evidence of life on Mars, while others have dismissed these worms as merely mineral artefacts (the pro-lifers are not claiming the worms arrived here alive). See, Barry Parker, *Alien Life: The Search for Extraterrestrials and Beyond* (Reading, Mass: Perseus, 1998), pp. 87-102, for a good discussion of the problem. Photographs of ALH84001 and its worms are included.

Stephen J. Dick traces the history of panspermia in the twentieth century in *The Biological Universe: The Twentieth-Century Extraterrestrial Life Debate and the Limits of Science* (Cambridge: Cambridge University Press, 1999), pp. 325-29, 367-77. The most prominent champion of panspermia in recent years was the Cambridge astronomer, Fred Hoyle.

partially cloud covered at all times. However, the clouds and the haze of the atmosphere do not present any serious hindrance to the observation of Earth. The satellites and such which surround our Earth present no problems at all to anything which attempts to observe the Earth from outside.

CARAS: Do you think that Earth has been contacted by radio or signalled? Do you think that signals have been sent to Earth?

DRAKE: It is possible that radio signals of extraterrestrial origin are arriving at the Earth at the present time with an intensity which is detectable with existing equipment. We cannot say for sure that this is happening and equally we cannot decide whether other civilisations are intentionally sending us such signals. There may be signals in other civilisations they use for their own purposes which are arriving at Earth. However, this is a very exciting and tantalising thought – to realise that at this very moment perhaps detectable radio signals from other civilisations are passing right through this room.

CARAS: What should we be doing that we are not doing?

DRAKE: If we believe that there are detectable signals and it is worth the expense of detecting them – and this is not a small expense – we should embark on a serious programme to search for such signals. Now, before we can have real confidence that such a search would be successful we perhaps need encouragement from other sources. Specifically it would be very encouraging to find that life existed elsewhere in the solar system. This would prove to us that our ideas about the origin of life and its great abundance in the universe are correct. Similarly it would be nice if we could observe at least one other planetary system to prove also that planetary systems are very abundant in space. If we knew both of these things for sure we could be quite confident that there are large numbers of intelligent civilisations and that given a thorough enough search the chances of detecting such civilisations by their radio signals, their own purposeful contact signals, would be very high.

SCENE TWO / TAKE ONE

CARAS: What will be the reaction on Earth? What kind of cultural shock would you anticipate if we suddenly did receive discernible artificial extraterrestrial radio signals?

DRAKE: I haven't got any real answer to that one. There would obviously be a great deal of excitement if we ever detected artificial extraterrestrial radio signals, but what the impact would be on the culture I think is hard to predict.

CARAS: Have we any room in any of our philosophies for it?

SK pausing for one of his occasional cigarettes inside the 'centrifuge' set

DRAKE: That's quite interesting. It turns out that some of the theologies have thought this one through. For instance, the Catholic Church is quite prepared to accept extraterrestrial beings. All that is required is that they have souls.

CARAS: Can you guarantee that?

DRAKE: That is something I cannot guarantee.

CARAS: What should we do – not speaking of cultural shock – but what should we do on the day that we receive a discernible extraterrestrial radio signal?

DRAKE: When we first detect an extraterrestrial radio signal – the first things we will do are the obvious scientific ones. That is you will analyse the signal, to see what form it is coded in. You attempt to see what kind of information is carried on it. It can very well be a signal the other civilisation uses for its own purposes and therefore it will carry some esoteric and perhaps uninteresting information. On the other hand it may be something that is designed for our consumption and will carry something that is of extreme interest to us… Now, the next step, of course, depends on just what the content of the signal is. If it is intended for us and conveys something important to us, a reply from us or questions perhaps would be called for. If it is on the other hand it is a signal which they are using for their own purposes, it is likely that the best we can do is just listen in on it and see what we can learn of this other civilisation from this signal.

SCENE TWO / TAKE TWO

CARAS: You said that the signal we conceivably would receive might be something for local consumption there and might be esoteric and uninteresting. Do you really believe that anything we found out about extraterrestrial life would be uninteresting? Or are we playing with semantics?

DRAKE: We're playing with semantics. Obviously any signal we find, even if it is for their own consumption and contains nothing more than the instructions of what passengers are travelling on what flight from here to there, would still be terribly interesting.

CARAS: Step number one is analysis. Step number two?

DRAKE: I've answered that.

CARAS: Yes. Step number three? Or is there a third step?

DRAKE: Well, step number three – if this civilisation is obviously trying to contact other civilisations we should answer. Now there are people who believe it might be dangerous to answer. Perhaps they will attack us? Perhaps we are the finest beef animals they have ever discovered? But the fact is, and this is something we should understand right now, is that we already transmit enough radio waves into space that we can be detected if they want to detect us. So, we cannot keep our presence secret. It is known if other civilisations really want to find us, so we might as well try to answer back. It is no use trying to avoid the issue.

SCENE TWO / TAKE THREE

CARAS: Is there a possibility, Doctor, that the radio signal that would come to us from an extraterrestrial source, an artificial signal, would be such that we could not or could never interpret it? Is this possible? Or is anything they could send us interpretable?

DRAKE: If we receive a signal from another civilisation that they are using for their own purposes it is possible that it will not be decipherable. If we receive a signal from another civilisation which is intended for our use it is virtually certain that we will be able to decipher it. Another civilisation will clearly devise a code which rather than being hard to decipher is easy. They will invent a code which is the opposite of the ones we are used to. The ones we are used to are the ones hard to decipher. These will be easy ones to decipher. And in fact we have now devised on Earth several codes which could be deciphered by another civilisation without any previous contact.

CARAS: What do you think the future is of the ultra-intelligent machine? Or the super-intelligent computer machine? Where is all this going?

DRAKE: I don't know anything about it.

CARAS: Do you know anything at all?

DRAKE: Can I add one comment?

CARAS: Yes, please.

DRAKE: One of the easiest codes to decipher is the equivalent to television. We have now constructed several examples of messages that one might receive from another civilisation which are designed as television pictures and it has turned out that some people on Earth have been able to decipher these in only a few minutes' time when they had the idea that it may be a television picture.

CARAS: Are these codes essentially mathematical?

DRAKE: These codes are very simple. They consist of a stream of perhaps two characters – dots and dashes or two tones or pulses – very simple messages in which the code consists of a very few symbols. Just one or two. Nevertheless, such a code is able to convey with ease a very complicated picture and thereby a very complicated message. And yet all of it is easily deciphered.

CARAS: Before, you did not want to speculate on what civilisation might look like within a few million years on top of the ten thousand that we roughly have. Is this an area you do not want to talk on at all? Or is there anything…

DRAKE: I think I would be very dishonest to pretend I could make any kind of reasonable projection because obviously a person a hundred years ago even could not have projected our present-day civilisation.

CARAS: A Neanderthal man plucked out of time and plonked down in the middle of a conference of scientists here in this time at the university – or exposed to Times Square – would be totally incapable of understanding things.

Faced with things none of which he has experience with – wheels, conversation the way we know it, electricity, artificial lights, et cetera.

DRAKE: That's our relation to them.

CARAS: Would we be possibly at that kind of a loss?

DRAKE: Yes. Well, it has been said that our relationship to a civilisation that might detect us is about the same as the relationship of a Neanderthal man to our civilisation. That is, we would understand and comprehend as little as would a Neanderthal man placed in one of our major cities today.

SCENE THREE / TAKE ONE

CARAS: What is the most logical means of communication either from Earth to extraterrestrials, or vice versa?

DRAKE: A lot of thought has been given to the problem of the most likely means of communication between civilisations. It seems very likely that the most usual means of communications will be radio signals and this is for one very simple but compelling reason. And that is that radio signals are far more economical than any other way we are aware of. In particular, they are far more economical than the use of rockets.

CARAS: What about laser?

DRAKE: Radio signals are far more economical than lasers for the simple reason that light comes in little packages and the amount of energy in each packet is dependent on the frequency of each transmission. A given package can only carry so much information. It turns out that a radio message consisting of radio frequency energy can carry about a million times as much information for a given amount of energy than any light signal. Therefore we think radio signals are far more likely than, say, laser signals.

CARAS: What about artefacts? Is there any likelihood that there exists somewhere on Earth something we haven't discovered that would give evidence of a visitation?

DRAKE: There is a possibility that there exists on Earth an artefact from an extraterrestrial visit. Now, we have estimated that at the very best – under the best conditions one would not expect a visit from another civilisation, an actual landing, more often than once in many thousands of years. Now, we would expect that such a visit would leave behind an artefact because otherwise no evidence of such a visit would be left to the intelligent beings that might come along later. So, such landings might leave artefacts. However, as you can see from the fact that they would come so infrequently the number of such artefacts would be very small and they would be buried perhaps now in archaeological ruins or what have you. And that only a great deal of excavation and good luck combined could lead us to such an artefact.

CARAS: Would we necessarily recognise them as of extraterrestrial origin?

DRAKE: One can't really be sure whether one would detect the extraterrestrial nature of such an artefact. However, another civilisation would probably go to great pains to make it clear that the object was of extraterrestrial origin. So I suspect that we would recognise it as extraterrestrial. We have to date found no artefact that could be construed as being of extraterrestrial origin. This is the prime evidence against the idea that UFOs, for instance, are spaceships from another civilisation.

CARAS: There have been cave drawings, and ancient writings or scribblings, picture writing that seem to show people in space helmets[2] of a sort. There's been quite a lot written about this. Quite a few pieces of this art reproduced. Do you give credence to any of it?

DRAKE: There are a number of examples of cave drawings, ancient scrolls, tablets and so forth which have very provocative drawings which appear in some cases to be drawings of creatures in spacesuits, pictures of space vehicles as we now know them and what have you, and people have wondered if perhaps these are records of a visit from an extraterrestrial civilisation. However, to date, no-one has been able to give convincing evidence that this is the proper explanation and, by the same token, in each case a completely natural, non-extraterrestrial explanation can be provided.

2 Caras is here alluding to the frescos from Tassili-n-ajjer in the central Sahara dating from around 6000 BC that show figures sporting ceremonial headgear that von Däniken and the 'Was God an Astronaut?' crowd have interpreted as space helmets and, therefore, evidence of extraterrestrial contact, no less. Two of the frescos are reproduced in I. S. Shklovskii and Carl Sagan's *Intelligent Life in the Universe* (San Francisco: Holden-Day, 1966), pp. 455-56. For an amusing chapter on von Däniken's nonsense by a scientist with an abiding interest in extraterrestrial life, see Ronald N. Bracewell's *The Galactic Club: Intelligent Life in Outer Space* (New York: W. W. Norton, 1976), pp. 97-104, 'The Chariots of von Däniken.'

Keir Dullea and SK inside the 'centrifuge' set

Frederick C. Durant III

[Assistant Director, Astronautics National Air and Space Museum, Smithsonian Institution, Washington DC]

SCENE ONE / TAKE ONE

CARAS: Fred, do you feel there is any possibility of our encountering evidence of an extraterrestrial intelligence in the years immediately ahead?

DURANT: Well, the possibility of discovering extraterrestrial intelligence is, of course, one that there is no answer to until the contact has been made. Certainly, I would expect that life of some sort, of some form, of some degree of development; perhaps much less than ours or much greater than ours, is a very definite possibility. One cannot predict when an encounter will happen.

CARAS: Do you believe it exists, personally?

DURANT: I do believe it exists. It is hard to imagine that life could exist on this planet only with the literally billions of stars and the billions of planets that we presume must revolve about some of them.

CARAS: Fred, is there any argument which militates against the possibility, or even probability, of extraterrestrial life?

DURANT: I certainly know of none.

CARAS: Do you feel it is a legitimate scientific pursuit at this stage of our own development? This search for it, actively?

DURANT: Absolutely! The search for extraterrestrial life is certainly one of the most exciting kinds of exploration. It will be done in a logical and scientific manner.

CARAS: What steps should we be taking?

DURANT: The steps that should be taken should be left to those who are biological scientists and are specialists in that field. Mine is in engineering and astronautics.

CARAS: Well, in the area of engineering and astronautics, what about radio monitoring? Do you think this is a legitimate pursuit? I am not asking you necessarily to give the scientific details of how this should be done, but do you think that we should be actively participating in radio observation programmes, passive or active?

DURANT: Both passive and active radio observation of frequency spectra appear to be the logical forms of exploration that are underway.

CARAS: Do you have any views of the origin of life on Earth or the origin of life on other planets?

DURANT: I do not feel competent to make a judgement on the origin of life either on this or on other planets.

CARAS: What do you think the reaction would be on the part of the human population on Earth should we – this is speculative – should we encounter direct evidences of an extraterrestrial intelligence? An extraterrestrial artificial radio source, say?

DURANT: I think the mind boggles at the impact of discovery by man of an extraterrestrial civilisation, or just extraterrestrial life in some form of development. Yet it is something which I think we will face some day. It may have the effect of bringing mankind closer together and national boundaries might then mean much less.

But, with regard to the impact upon society of such discoveries, I think that even today, the public underestimates greatly the impact which will occur when man first sets foot on to the Moon and commences extraterrestrial exploration. I believe it will have impact around the world that is still little appreciated.

CARAS: Will the impact be less than when man steps on Mars, because mankind will already be used to the idea? Or is each step that we take into the future bound to have enormous scope for repercussions?

DURANT: One can't say. But let's get to the moon first! We have a national programme to do that. The technical capability for a long journey to Mars will evolve from success of the current Apollo programme.

CARAS: There has been an awful lot of talk in years past and there has been an awful lot published, even recently, trying to demonstrate that Earth has been visited in the past. This ranges from ancient launching pads to drawings in caves thirty-five thousand years old showing men in space helmets. There has been a lot said and a lot published about this. Do you give any credence to any of it?

DURANT: The possibility that the Earth has been visited by extraterrestrial life some time in the past is simply unknown. It would be ridiculous to suspect that it has not occurred at some time, hundreds or thousands of years ago. However, no extraterrestrial artefacts have been identified to date.

CARAS: Do you think there is any evidence yet that radio signals reaching Earth from extraterrestrial origins are artificially produced?

DURANT: I am not expert in such matters. However, I know of no reports of signals where there were signals that have yet been received which can be construed as coming from intelligent beings.

CARAS: If we stretch it to the breaking point, we can say that civilisation on Earth is ten thousand years old and Earth is not a very old planet and our Sun is not a very old sun. Let us extrapolate or imagine a little bit. What might intelligence be like if it had four million and ten thousand years to develop, instead of ten thousand years? What would they be like? What would their capabilities be from the point of engineering science?

DURANT: The problem of conceiving the level of capability, technology and the development of life which is millions of years old, or intelligent life that has been intelligent for millions of years, simply staggers one's imagination. It is difficult even to begin to conceive of the degree of communication that might occur; and to learn the interests, pleasures, if you will, of such life. One simply cannot conceive of it, but we have a much closer problem really in trying to conceive today of what will happen, right here on earth, in the next ten years, let alone twenty, thirty or forty years. In my mind the general public has quickly forgotten that less than ten years ago we in the U. S. were attempting to place twelve pounds in satellite orbit. Today eighteen tons of payload can be lifted into orbit by the Saturn booster. Through proven rendezvous techniques large permanent space stations can, and I'm certain, will be constructed. The rate of current development of technology is generally not appreciated. By the same token, I am convinced that the rate of development of technology will be even more rapid in the future. I conceive of man's current rate of progress as being almost exponential.

SCENE FOUR / TAKE ONE

CARAS: What might some of the advantages be to mankind should he be able to make contact with an extraterrestrial intelligence very much in advance of his own?

DURANT: Should man make contact with extraterrestrial intelligence very much more advanced than our own it would be hoped that some type of friendly relations might be established. The next step would be, it seems to me, to attempt to learn from such intelligence, presuming communication was possible. If it was too far in advance, of course, one just doesn't know, it might be that mankind might not even be considered important to the alien visitor.

CARAS: You said that it would be hoped that a friendly relationship would be set up. Is it the fear on the part of some people, mainly science fiction writers, that it would be an unfriendly relationship? Isn't this simply projecting man's own weakness? Isn't it to be assumed that if a civilisation were to be very much more ancient than our own they wouldn't have the animosities, the hostilities, the insecurities that make us behave as we do?

DURANT: It could be presumed that an advanced intelligence or race of some civilisation would have learned in the eons of their own progress how to get along, how to meet and mix with 'foreigners' of their own variety if not those from some other planet. Presumably that by some means travel was possible. I think the thought of or even fear as projected by some science fiction writers that such races of advanced civilisation that man might make contact with, the fear element, is, perhaps, really a projection of man's own fears and worries in today's world.

CARAS: Machine intelligence we talk of, where is it going, Fred? What is going to happen in the next ten years and beyond that? What can we look forward to machine intelligence doing for us?

DURANT: The possibilities of utilising machine intelligence, advanced computer systems, with great memory circuits, is a possibility in the future. Surely, as man develops more complex machine intelligence he will learn to teach such things how to perform routine acts, how to make simple judgements. This is as far as one can predict based on today's technology. It would be foolish to think that as man develops machine intelligence, more and more complex decisions, eventually perhaps approaching man's current level, would be expected to follow. It is possible to imagine even that more rational decisions might be made by a machine eventually than by man. This brings up the interesting possibility of turning over certain state decisions to the machines. However, this again is more than a decade away which I think is the limit that anyone can predict these days.

CARAS: There are some scientists who feel that there is a distinct limitation on electronic machinery. That the electronic computer, as we know it, is a dead-end device and that the super-intelligent machine, if you like, of the future will be a machine that is a bios, a biological machine. What do you feel about the bios versus the electronic machine?

DURANT: Who knows the limitations to the electronic machine approach? I do not see that limits can be established with fast computer systems – and fast being a relative term – memory circuits, and memory again being a relative term, this doesn't mean necessarily that there is a limit on their development. Certainly bio-sensing techniques will be developed, I believe, hand in hand, so that eventually a combination of the two may be used. But this is something we must wait to see.

CARAS: These machines get smaller and now we can assume that they will. Do you think we could ever relinquish certain things to them? For instance, genetic advice, genetic control? Do you think we could relinquish this kind of decision to machines?

DURANT: With respect to the kinds of decisions that can eventually be left to machines, it is difficult to imagine that decisions relating to future generations such as genetic selection techniques would be left to machines. One cannot know. But it would seem to me that this is playing with many factors that are unknown to us at this point. But I would tend to let nature and natural selection take precedence at this time.

CARAS: Do you think the time will ever come when there will be a difficulty in distinguishing between man and machine? Do you think that machines could become sentient in the sense that they will feel pain or have emotions? I'm talking now of far, far into the future.

DURANT: I can imagine the possibilities that advanced forms of machinery at some time in the future might feel pain, pleasure, have emotional responses if it was decided and found that building in circuitry as counterparts of man's own experience and emotions, assisted in future determinations that such machines might make, that such emotions and sensory perceptions might be desirable and might actually occur.

CARAS: Do you think there is any possibility of a machine, if it is a super-intelligent machine, whether it is a bios or whether it is an electronic machine, could it have personality disorders? Would you get a machine that requires psychiatry? Would you have one machine put another machine on a couch and analyse him?

DURANT: Certainly if man develops machines that have sensory perception and emotional responses that, if you will, feel pain and pleasure – if in any way they appear to respond as man does then certainly we can expect them to become upset or have neuroses as man does today. Certainly I could imagine that certain machines eventually might require rest periods and an opportunity to reconsider their emotions. What we're thinking about really is an attempt to create something very much like man. I think there is another approach in attempting to build something that does not have perhaps the weakness of man.

SCENE FIVE / TAKE ONE

CARAS: Fred, the past, present and future of technology and space travel. Why don't we have just a few minutes of generalised remarks about the history and what has happened and what has surprised us and what we did not anticipate, and some kind of rough timetable? We are not going to hold you to this. If we do not hit Saturn in 1994 we will not come round and look you up! A generalised view and review of a look into the future

DURANT: In ninety seconds!? Man has always wanted to explore space from the moment it was recognised that the Moon, planets and stars are actual bodies a long way off. The 'grass is always greener over there' philosophy has undoubtedly played a role. It is human nature to want to explore and, once man has obtained the tools to do this, he is going to do it and is doing it. There is always the challenge of taking risks. Without some risk there is no progress. Space flight was a dream until the last decade, until the use of old concepts, the rocket motor, the development of technology that came with the aviation industry, and high-speed aircraft, and the long-range ballistic missiles became the springboard on which man has now leapt into space. From the standpoint of history, we are living in an absolutely explosive growth of technology. The rate of progress is beyond anything that could have been predicted, five let alone ten years ago. The expenditure of seven billion dollars each year by the United States, quite apart from what the USSR is spending in their developments, cannot help but engender many new advancing frontiers moving in ever widening circles, so there are literally hundreds, thousands, of new technologies being developed. The cumulative effect of these technologies as mankind moves into space, as he makes new discoveries and gains better understanding of nature, he is moving on to new levels, new plateaux of capability. How fast we will move, beyond the current programme of planned exploration of the Moon, one simply cannot judge now. We are told that if we need five, or eight or even ten years to plan for a trip to Mars this may be so now but, in my opinion, five years from now we will be able to plan such a journey. Beyond Mars, certainly we will have unmanned exploration of Venus and, if it seems feasible, manned exploration on the surface, despite the temperatures. Eventually, visits to at least the moons of the giant gaseous planets will come about.

CARAS: What is the ceiling on man's capabilities? Where will man stop? Or will he be stopped?

DURANT: I don't believe that there are any ceilings or limits to man's capability for exploration. In other words, I do not believe man will ever stop exploring space. He will go farther, longer even perhaps with colony transportation, reproducing and with whole new generations, exploring the universe. I do not accept that there is a limitation to travelling at the speed of light only. Of course, our current capability is such a tiny fraction of the speed of light.

We do not know the future. However, one cannot accept that there will never be discovered ways of travelling faster than the speed of light. I have not the foggiest notion of how it will be done. We know of no way to negate the force of gravity and yet, in my opinion, the possibility exists. I do not have the slightest notion how it might be accomplished. But, if and when it should be accomplished, it will appear just as miraculous and magical to us today, as would perhaps a television or solid state physics to Thomas Edison. There are technologies being developed today that may result in great advancement of understanding. As we sit here today we cannot envision the accomplishments of the next decade, and beyond.

CARAS: One last question, Fred. Is there anything, do you believe that you could imagine man doing, that man would one day not be capable of doing? Is there anything within our imagination that man could not do eventually?

DURANT: I do not foresee any limitations to man's technical capabilities. One must face the problem, however, that he can destroy himself and all of civilisation. Rocket power itself is like the two-headed god, Janus, which faced towards peaceful endeavour, as well as towards hostility, warfare and destruction. It is up to mankind and society to make the decision how to use newly discovered sources of energy and their technological power.

CARAS: Anything to add, Fred?

DURANT: No.

LOOKING BACK: Frederick C. Durant III, 2003

So! What changes seem necessary or desirable to the transcript responses to Roger Caras' questions thirty-seven years ago? First, let's delete my repetition or rephrasing of the question while I mulled over how to respond (Done!). I am bemused by the air of confidence and enthusiasm of some of my responses. Ah, the assurance of youth, bordering on arrogance! Still, I find no judgments to which I take strong objection today.

Does this mean that no progress has been made in understanding of nature, physical principles or cosmology? Not at all. Data, a few months ago from the WMAP (Wilkinson Microwave Anisotropy Probe), a million miles from earth, has refined statistics of the age of the universe (13.7 billion years) and the appearance of the first stars 200 million years after the Big Bang. And, as so often happens, an outline of greater scientific puzzles arises: four percent of the mass of the universe is now calculated to be 'matter,' twenty-three percent unseen 'dark matter,' and seventy-three percent 'dark energy'! Theoretical physics will chew on these implausible data and their meanings for years. The import? Unknown and perhaps incomprehensible in our lifetime.

Do these newly refined statistics have import to the search for extraterrestrial life? Perhaps, but outside my cozier field of engineering, studied sixty years ago.

Fifty years ago I was fortunate to be involved in a brief study and national security evaluation of UFO phenomena. The findings and conclusion reached by this select group of world-class physicists has stood the test of time. No data nor factual evidence has appeared in the intervening years to alter that report.[1]

The speed of digital computers has increased 'big time' but as for the meshing with bio-systems and DNA, I plead ignorance and leave to those working in that potential field.

As to the motivation, risk and excitement of manned exploration of space? I see no change. Mankind will surely provide individuals competent, willing and eager to risk their reputations and lives in such quests.

SK's planning centre for the SFX shots

1 The famous 'Durant Report,' more properly *Report of Meetings of the Scientific Advisory Panel on Unidentified Flying Objects*, convened by the Office of Scientific Intelligence, Central Intelligence Agency, January 14-18, 1953. There are several postings of it on the Web. (AF)

Freeman J. Dyson

[The Institute for Advanced Study, Princeton, New Jersey]

6 APRIL 1966 AT MGM STUDIOS, BOREHAM WOOD

CARAS: Doctor, what do you think the likelihood is of our encountering extraterrestrial intelligence in the next few decades?

DYSON: Well, I'm very puzzled I must say about the situation here. I don't think I feel it's very likely. It is, of course, impossible to judge. I'm sure we ought to be keeping our eyes open for it and I'm sure we should be looking harder than we are to try and find something. On the other hand, it's sort of puzzling that we haven't seen anything yet if it is there because, if you actually think about what an intelligent species would be likely to do when it's got millions of years beyond the point which we're at, it would be likely to look very conspicuous to say the least of it.

CARAS: Do you feel convinced there is extraterrestrial intelligence – aside from the idea of contacting it – but do you feel that there is such a thing?

DYSON: Well, that's what I say. I'm very sceptical of it because I have the feeling if it were there it would probably be glaringly obvious, but nevertheless, I hope it is there.

CARAS: What are your views on the origin of life on Earth?

DYSON: Well, this of course, I only take from the biologists. I'm not an expert in that field myself but I think it's presumably true as the biologists now believe that life can originate by gradual development from something that starts off as pea soup and ends up as little animals and that this is in some sense a natural process.

CARAS: Do you think there's anything in the idea that life may have come to us from cosmic dust? That it didn't originate on Earth at all?

DYSON: Well, it's always possible. On the other hand the conditions that this dust would have to survive are pretty rugged and it's certainly not likely that any spores which we now know about would survive such a trip.

CARAS: Well, what do you think are the most realistic, most reasonable arguments in support of extraterrestrial life, admitting that they don't necessarily give us a conclusion? What do you think are the best arguments for and the best arguments against?

DYSON: Well, the argument for is very simple. It just says they're not all that important in the scheme of things. They're just as it was thought at first. The Sun revolves around the Earth and now we know that just as we're one of many objects revolving round the Sun and then it was discovered the Sun was only one of very many objects revolving around a galaxy and so it seems just from point of view of our modest place in the overall scheme of things it seems very unlikely that we are the only intelligent species in existence. But this is, of course, an argument which has no scientific value. It is purely a hunch which may be convincing to some people and not to others.

CARAS: The future of machine intelligence, how far do you think it is going?

DYSON: I don't believe that electronic machinery is going to go very much further than it is now. This is, of course, something that we'll find out in the course of time. I may be completely wrong, but there seem to be some kind of natural limitations on what one can do with electronics, which fall very far short of what the human mind can do, so I would not believe that anybody with existing types of electronic machinery could build anything that resembles a human mind. But, on the other hand, I would say when we learn how to use biological techniques ourselves and to build machines with biological materials then it will be very different and then it may well be possible to go far beyond anything we have at the moment and we shall probably be able to create intelligence and whether we will call that machine intelligence or not is, of course, a matter of words. I think it will not look like an electronic computer but it will look much more like a living organism.

CARAS: Will we be able to finalise the distinction between man and biological machinery?

DYSON: Yes. Ultimately, I'm sure it is bound to.

CARAS: Is there any particular reason why we would create a machine in our own physical image?

DYSON: I would say there is no particular advantage but, of course, one can't predict what people will want to do. I would certainly not expect that.

CARAS: Are we on the threshold of this breakthrough into biological machinery?

DYSON: I would think so, but I'm not a biologist. I know the biologists are in general very optimistic or pessimistic concerning the things they are going to do. They believe they are on the threshold of great things. I'm not able to judge independently, but I would think it's very likely.

CARAS: [?].

DYSON: Well, this is what puzzles me. I can't see any limits to what we could do. It seems to me clear that we could turn the galaxy upside down if we wanted to within a million years and there's nothing in the world of physics at least that would prevent us from doing that. There may be some good reasons for not doing it and there may be good reasons why other intelligent species are not doing it, but all we know for sure is that it's not being done.

CARAS: If there were an extraterrestrial intelligence with an age far greater than our own, might they not just want to stand back and look at us?

DYSON: That may well be true but what puzzles me is why we have not seen them, quite apart from the question of communication. I would have expected if there existed any highly developed race of technological beings that looking for them would be like looking for evidence of life on Manhattan Island. That they would in fact have just transformed the whole surroundings completely. But, evidently they haven't done this for some reason, if they exist.

CARAS: After all, it's what, only a couple of hundred years that we would have known what we were looking at, if we were looking at them? If they had landed even a hundred years ago?

DYSON: No reason, except the fact that we don't see any traces of them when we look in the sky, and this is peculiar. If an Indian were to come into New York Harbor he might not understand what he sees but he would at least notice there is something there.

CARAS: Going back for a second to what you were talking about earlier [?].

DYSON: Well, this, of course, is a question of whether the name will mean the same thing. I think it could.

CARAS: Do you think this is the key or a gateway or an opening gambit for biological immortality?

DYSON: In certain sense, yes. But I would say there are much more drastic problems involved than tinkering around with human bodies and that if you are tinkering around with the human body you inevitably tinker with the mind as well and where this ends I certainly can't foresee but I can see clearly that there is a lot of trouble ahead.

CARAS: [?].

DYSON: Well, we simply don't know enough to say. I don't think even the most ardent psychoanalyst would claim that he really understands neurosis and so I don't see how you can possibly say in advance what would happen.

CARAS: [?].

DYSON: One hopes so. I mean, honestly, if you look at the way our brains operate we are subject to all kinds of problems merely because we can't remember things. We can't learn things fast and efficiently. There are many practical limitations that we're subject to which one imagines one could overcome if one knows how to do it. So one could hope to improve our brains to the point where we can learn many languages and understand everything that is known in science. In general, be much more competent men to live than we are at present. And one hopes one could do this without radically changing our moral standards and our personalities. But, of course, it's very likely that all kinds of unforeseen effects would appear at the same time and, perhaps, including new forms of neurosis. So, it's a game which one has to play extremely carefully, but I have no doubt whatever it's going to be played and in some ways I'm excited at the prospect and in other ways afraid.

CARAS: [?].

DYSON: In the next thirty years?

CARAS: Yes, the next thirty, thirty-five years.

DYSON: Yes, I'm sure it is a medical development but I don't know in just what direction, but I would guess we shall be in some sense controlling our own heredity by that time and there are such obvious advantages in doing this if one could be sure when one has a baby that it will not be deformed and not be mentally deficient, one would certainly be only too happy to take advantage of whatever science can do, but once you have these powers then, of course, there are other things you can do besides merely eliminating the harmful effects and so we are going to come up against these very serious questions as to who is to decide what kind of children are to be born. This may or may not happen in the next thirty-five years, it's hard to judge. But it's something which could happen.

CARAS: [?].

DYSON: Well, you ask many questions all at once. In the first place we do understand gravity. That is perhaps an immodest statement for a physicist to make, but we understand gravity somewhat better than we understand anything else in the physical universe. I think this is a true statement. We understand it in the sense that we know precisely the laws which it obeys. We understand very clearly all experimental manifestations of it that we observe – and this is hardly true of other physical effects, other forces in the universe. So, in a certain sense, gravity is the best understood of all the physical forces that we know about. And this has the consequence that there is the least to be hoped from gravity from the point of view of practical applications since it's already so well understood. It's fairly clear what we can do and what we cannot do with it and any idea of switching it on or off, I would say, was out of the question.

CARAS: [?].

DYSON: This would be reasonable, but still the Earth gravity is not so very strong either. There is no reason for not taking off large ships from the Earth.

CARAS: What about the famous anti-gravity ships in science fiction?

DYSON: I don't believe that this will ever happen. Of course, that is a strong statement but there are some parts of physics which seem to be very solid and I would say this is one of them.

CARAS: Some physicists have suggested that some sub-nuclear particles are actually other sub-nuclear particles moving backwards in time?

DYSON: I don't believe this makes any sense. The flow of time is again something that is fairly well understood in the physical sense and it's a consequence, I mean, the fact that time has a definite direction of flow is in some ways a consequence of the fact that we are living beings and that we take in information and make use of it and that this is done in accordance with certain definite physical and mathematical principles. I don't believe this has much to do with fundamental particles, the fact that you can talk in a sensible way about particles travelling backwards in time. It does not mean that we should ever be able to travel backwards in time or that we could imagine other beings to travel backwards in time and this is, in certain ways, just a confusion of the language.

CARAS: What about time itself moving backwards?

DYSON: I don't know what that means. I'm not inclined to take it seriously.

CARAS: [?].

DYSON: Well, the greatest area of ignorance is certainly in the area of very high energies where most of the exciting developments in physics are being made at the moment. This is the world of small particles crashing into each other at very high energies and producing other particles. All these things are not at all well understood and are very exciting to scientists just for that reason and that I should say is the main unexplored area. There are other unexplored areas. For example, wherever you look there are exciting problems which are not solved. For example, why does the Earth have a magnetic field? Why does the Sun have spots? Why do the planets all go around the Sun in the same direction? There are innumerable questions of this kind which have not yet been answered. These are all interesting areas, but I should say the 'Great Unknown' is the world of high energies.

CARAS: The fact that the planets seem to be so well ordered round the Sun seems to suggest that there are basic laws?

DYSON: This is not logically necessary, but it's certainly very likely and actually there have been seen kinks in the motions of several nearby stars, indicating they do have planets. There's one particularly clear example of this where it's well established that there is a planet of about the size of Jupiter going around another star.

CARAS: Is there anything observable from our solar system that makes it appear to be in any way unique whatsoever?

DYSON: Not at all. In fact, rather the contrary. That everything we know about the solar system indicates that it's a very ordinary type of object.

CARAS: How many times do you think it might repeat itself?

DYSON: Well, this is something one can only guess but if the guesses are right then there are something like one billion (the American billion, that is to say) such objects in our galaxy alone.

CARAS: [?].

DYSON: All right. Say there are in our galaxy something like a billion stars which so far as we know would be suitable for a planet to exist in and each of which might have five or ten planets. Whether these actually are there, of course, we have no means of knowing.

CARAS: So that there's nothing observable from the Earth and its fellow or sister planets that makes us in any way unique?

DYSON: Nothing whatever except for us.

CARAS: Are there any chemicals or physical [?] of our planet itself in the solar system?

DYSON: Not as far as we know, but we know very little about any of the other planets and even less about any other planetary systems, so one can't say. But everything we have found out up to now would support the view that we are in quite an average place and that there may be many others like it.

CARAS: [?].

DYSON: I would say more than that. It's not only legitimate but it's an extremely important part of astronomy. I think it's not only legitimate to start looking for signs of intelligence in the universe but it's also extremely important and it's something which should be done much more intensively than it ever has been done. Up to now we have only looked for about a month at a couple of stars and that is all and nothing more has been done. So I hope very much that this kind of search will be revived and made into a routine operation and that it will be taken more and more seriously as time goes on. It clearly is an extremely important question to decide whether or not there are other intelligences in the universe and it would have profound consequences if one found any positive evidence. Even though I'm very sceptical as to whether we shall find positive evidence, I'm all the more anxious that we should try.

CARAS: [?].

DYSON: This is, of course, the whole charm of the subject – that it's impossible to foresee what the consequences would be which we have absolutely no way of imagining. I don't remember who it was who said that any alien form of life would not only be stranger than we imagine but it would be stranger than we can imagine, and this is profoundly true. So, what would be the consequences we can't imagine either. All one could say is that merely even if we had no communication but merely observed passively what was going on we would probably see all kinds of things that are totally unexpected and different from things which we see normally and this would certainly make big changes in our picture of things, probably big changes in our view of ourselves, in our view of our place in the world in general.

CARAS: [?].

DYSON: Very few. One is the velocity of light and I don't believe that anything will ever go faster than light and this is a basic limitation. It means that travel over distances from one star to another will always be very slow, will always take a long time, and therefore it will not be like going to visit one's family in the next town. This limitation of the velocity of light is, I believe, quite fundamental and I'm happy about it. I think it makes space travel much more interesting that one can go to places which are remote, not only in space but also in time. But I believe there are very few other limitations. It's not difficult to imagine methods of propelling oneself with nuclear energy which go at some fraction of the speed of light, some few percent perhaps of the speed of light, and I believe in time one probably will get up fairly close to the speed of light, but this we don't understand yet how to do.

CARAS: [?].

DYSON: Oh, sure, sure. In fact, the solar system as a whole is very accessible. I would think we could probably within a hundred years. We could probably be travelling around the solar system fairly freely. Because if you take nuclear energy and, for example, a system of propulsion using electricity and nuclear energy generators and plasma jets and things of this sort one can imagine fairly efficient and economical transportation around the solar system within a hundred years. Anything beyond that, of course, is a vastly more difficult proposition and would certainly take more than a hundred years.

CARAS: Do you think we will be able to colonise any bodies in our solar system?

DYSON: I think that one of the interesting questions is whether we can make colonies in the solar system within the foreseeable future, whether the human race is going to spread out in many different parts of the solar system. I believe we shall but, of course, I have no means of telling where or how. We don't yet know enough about the places that might be suitable any more than Columbus knew what he was getting into when he sailed out from Spain. But one place which I think looks very promising is the comets, and most science fiction writers concentrate on planets but, in fact, it may be that the most interesting places for life to settle are in fact the comets. The point being that in the solar system there are ten planets and there are probably several millions or several billions of comets. A comet might be very suitable as a place to start a colony. It is made of materials which are familiar to us: water, carbon, nitrogen and oxygen, which are just the materials one needs for living plants and animals, and the comets have the great advantage that there are a great many of them and it would probably be easy to establish small colonies given a certain degree of technical development which we don't yet have.

Selecting a camera set-up with a Panavision viewfinder
over Gary Lockwood's shoulder

Gerald Feinberg

[Professor of Physics, Columbia University]

9 MAY 1966 / ROLL ONE / TAKE ONE

CARAS: Doctor, will men ever travel to the stars?

FEINBERG: I think men will travel to the stars sooner or later. The factors that will decide that are that some men want to, and that even today we can think of ways in principle by which it can be accomplished. Some scientists have said that the stars are too far away for us to ever reach them, but distance by itself doesn't mean anything. In fact, distance is equal to speed multiplied by time, and if relativity and the amount of energy available keep us from travelling too fast, we can always try to extend the amount of time we have available. There are various ways we could do this. One is by extending the human life span, which would be a good idea for other reasons also. Another would be by some kind of suspended animation for the travellers, either by lowering the body temperature, or by some kind of chemicals which decrease the rate of metabolism. Already some scientists are working on methods like this. So I think that sooner or later it is pretty likely that some men from Earth will travel to other planetary systems.

CARAS: Do you believe there is intelligent life in the universe – other than on this planet?

FEINBERG: It's hard to be sure whether there is intelligent life elsewhere in the universe, because we don't even know if there's any life at all – there's no direct evidence for that right now. But if we're willing to believe that there is some kind of life elsewhere in the universe, then I think it's reasonably likely that some of it is intelligent. The history of science has several times involved a change in the attitude that the Earth is somehow special in one way or another, and I think that that's likely to happen again in connection with intelligent life on other planets. If there is life elsewhere, then judging by the fact that we've progressed from, say, the invention to our present stage in about five thousand years and that in the next five thousand years we're likely to progress much further, I think it's pretty likely that life on other planets, somewhere, may be much more advanced than we are now, because it may have gotten started five thousand, or for that matter, five million years earlier. So, somewhere else in the universe, there may be life which has already reached the stage that we're not going to reach for the next five thousand or five million years on Earth.

CARAS: Can you conceive of life originating elsewhere and not going into an evolutionary cycle. Is not intelligence the ultimate achievement that all life is destined to realise?

FEINBERG: Well, one likes to think that. It's hard to know whether it's true or not, but my guess would be that it is.

CARAS: Speculating for a second, what might some of the capabilities be of an advanced civilisation? Let's just take a million years. Let us say, for instance, that there was intelligent life and it was a million years older than ours, what might some of its capabilities be?

FEINBERG: Well, I think that in a million years the human race would be able to do anything one can think of right now that doesn't specifically violate the laws of nature and, perhaps, many things that we think of as violating the laws of nature. So, by extension, I would say that some other form of intelligent life, which is already a million years ahead of us, would be able to do all of those different things.

CARAS: What might some of those things be? Could they change the orbits of planets? Could they re-order their planet? Could they create usable quantities of water out of rock? Would their body change? What might some of these enormous changes be?

FEINBERG: Well, I think some of the things that such an intelligent race might do would be astronomical changes, like making their planets move differently or, perhaps, even arranging them into some kind of geometrical array which had some value for whatever their purposes were. I think they would also do a great deal to reshape their own biology. Even if a race is much more intelligent than we are, it's unlikely that they would have reached any kind of perfection: so one of the things I foresee an intelligent race wanting to do is to make itself even better than it is. I think we'll do that to ourselves sooner or later and I expect that another race would do the same thing.

CARAS: What are your views on the origin of life on Earth or elsewhere?

FEINBERG: I don't have any real ideas about that – it's not my speciality, so to speak.

CARAS: If there is intelligence in the universe, and we're working on the basis that there is and they have these advanced capabilities, do you suppose they would have visited us in the past?

FEINBERG: Well, that's a very good question. It is a little surprising if there are super-intelligent races around, why we haven't seen some indication of them.

One possibility is that they had visited us, say half a million years ago, when there probably wasn't anybody around to know that we'd been visited. I myself don't think it's too likely that they're visiting us every day now, which some people believe is the explanation of the flying saucers. But the Earth has been around for a very long time and man, at least in his present form, has been here only one ten-thousandth or so of the life of the Earth, so that the chances that we would have been visited while we were able to recognise it is pretty small.

CARAS: What do you think about the investigation into extraterrestrial radio sources? Do you think that any of these show any evidence of being artificially produced?

FEINBERG: I don't think any of the radio sources that have been detected so far are artificially produced. I think one reason for that is the amount of energy necessary in order for us to be able to detect a radio source is very large – probably so large that even a super-intelligent race would have trouble controlling it. I think we have to look very carefully to find specific examples of someone broadcasting to us. And that's been done only for a very short time, so it seems to me unlikely that the radio sources that have been detected are examples of intelligent life. On the other hand, it may be that I'm too pessimistic and that the possibilities for controlling energy are almost unlimited, so that even something which pours out as much energy as a quasar is some indication of conscious control.

CARAS: Back to what we were talking about before. Doctor, do you think that the continued search for extraterrestrial radio sources is a legitimate scientific pursuit?

FEINBERG: Yes, I think that it's a very good idea to keep looking for some signs of extraterrestrial life. For one thing it would have a tremendous effect on our civilisation, I believe, if we did find such an indication – we'd sort of feel much less alone in the world, I think. Also, it's possible that if we could set up a real communication with extraterrestrial beings, we'd find out the answers to some of the questions that we'd very much like to know – like how to control ageing, or things like that. I think it may be a little too hard to communicate with extraterrestrial beings, even if we do find evidence of their existence. I'm not sure that we and they would think alike – but I think it's worth trying at least.

CARAS: What sort of language would we use to communicate with them?

FEINBERG: Well, that's a good question. I'm not sure. There have been efforts made to develop artificial languages which would, supposedly, make it possible for two groups, neither of which understands the other's thought processes, to eventually get together – but I remain to be convinced that that would even work between two different human beings, let alone between one race and ours.

CARAS: What do you see as the ultimate development of computer machine intelligence?

FEINBERG: Well, I think that in the very near future – say, within twenty or thirty years – we're going to develop computers which are first as smart as we are, and then a lot smarter. The important idea to keep in mind in this respect is that anything that nerve cells can do, can in principle be done better by transistors and their descendants. Computer technology is only about twenty-five or thirty years old and already it is possible to make computers do such specifically human things as playing chequers, or solving geometry theorems. In fact, I saw a geometry theorem that was proven by a computer, which the man who programmed the computer couldn't have thought of, and which if a high-school student had invented you would have predicted a great future for him in mathematics. I think that developments like that are pretty likely to continue in the future and that fairly soon the computers are going to think better and faster than we can. But I don't think that the future is necessarily bleak for human thought either. We're just beginning to get some idea of how we can manipulate the genetic pattern to influence the functions of living cells. If we can do that, then I think it should be possible to produce human beings who can also think better and faster than we can. In fact, I think it's a very exciting prospect that fairly soon there'll be some kind of intellectual relationship between men and machines in which each of us contributes those aspects of thought that we can do best. I think this is something very much on the cards and something which I myself am looking forward to with a great deal of interest.

CARAS: Freeman Dyson said recently in a conversation that he thought that the electronic machine had very, very limited capabilities and very limited potential and that the real potential lay in a biological machine. And other men have said this as well. Do you feel that the super-intelligent machine that we shall develop will be a bios, an electronic machine, or what shall it be?

FEINBERG: Well, I think we'll develop both. I think that the possibilities inherent in biological systems are nowhere near tapped completely and I think that through genetic manipulation or, perhaps, conscious redesign, we can produce creatures that think better than we do now. But I don't think that the outlook for electronic machines is so small either. There are many advantages that electronic machines have over biological systems. One is that the rate at which information flows within them is many, many times faster within a machine, an electronic machine. The information flows at near the speed of light. Within human beings the nerve impulses flow at a few hundred feet a second which is many, many orders of magnitude slower. I think that difference in itself gives the machines a very great advantage and I see no reason why that advantage can't be put to use. But I think that both developments will come.

CARAS: When the biological machine does come and then begins its advance and its march towards perfection which it soon will inevitably achieve, will the point not come when there will be trouble in distinguishing between man and machine? Will there be a point when we have a kind of fuzzy boundary?

FEINBERG: I think there may be some trouble in deciding when a machine is conscious. An intelligent machine wouldn't necessarily think the same way that human beings do. In fact, I think it's very unlikely. I believe we probably will make intelligent machines, either biological or electronic, before we've understood precisely what intelligence is in human beings. But the fact that two systems don't operate the same way doesn't mean that you can't recognise some kind of functional relation between what they do. What criterion we will use for deciding that a particular creature, if I can call it that, is conscious I don't know. I think there are going to be some interesting philosophical questions involved in making that choice.

ROLL TWO / TAKE ONE

CARAS: When do you envision, in time, the appearance of a super-intelligent machine of the magnitude we were discussing?

FEINBERG: It's hard to predict very accurately when it's going to happen that we get super-intelligent machines. But I believe that it will come sooner rather than later. I think within the working lifetimes of students now coming out of the universities this will have happened – so, say by the end of the 20th century – or in particular in the year 2001.

CARAS: How will such a computer or such a machine be programmed? Will it be run and wired and sent to school or will it be programmed as today's primitive devices are?

FEINBERG: I don't think I can answer that question. I don't know enough about it.

CARAS: When the machine comes – whether it is a bios or whether it is a pure electronic device, and it has its magnitude of intelligence – what about neurosis? Will it develop personality characteristics, both adverse and beneficial?

FEINBERG: It will probably develop some kind of personality characteristics. I'd hate to have to predict what they are going to be. Human psychology is a hard enough subject. Robot psychology I think is right now an impossible subject to speculate about. But I suspect that it may be easier to treat the mental diseases of machines than of people.

CARAS: If they are as bright as we hope they will be, wouldn't they be able to sort themselves out and couldn't they put themselves on a couch and sort it out?

FEINBERG: Yes. You might have one computer acting as psychiatrist to another one.

CARAS: Is that a bizarre idea?

FEINBERG: Well, no more bizarre than some of the other things that are going to come about.

CARAS: Will we ever have an emotional responsibility or an emotional interaction with a machine that achieves this magnitude of intelligence?

FEINBERG: I would guess that there will be emotional relations between men and machines run in connection with what might be called a generalisation of technological unemployment. There has been a lot of talk about how machines are putting people out of work – but people who talk about that usually have in mind white collar workers or such – but the super-intelligent machines are going to put everybody out of work, even the people who are the most creative, like scientists or, for that matter, artists and so on. Sooner or later the machines are going to be able to do, say, theoretical physics better than any human being and at that stage the theoretical physicists who right now are the only ones who can do that kind of science are going to feel the same way as some of us who have been put out of work by a pile driver like John Henry. That, I think, will be a fairly serious problem for the theoretical physicist, but it will be one that they will have shared with a lot of other people before.

ROLL THREE / TAKE ONE

CARAS: Doctor I J Good up at Trinity, Oxford, wrote a paper once that the super-intelligent machine, which was a necessity for man's future, would be the last invention man would need to make.[1] Have you any comments to give about that?

---

1 'Speculations Concerning the First Ultra-Intelligent Machine,' in *Advances in Computers*, vol. 6, pp. 31-88, 1965 (Academic Press).

FEINBERG: Well, I would hope that the super-intelligent machine isn't literally Mankind's last invention. Some people have worried that we might invent these machines, and then they might decide that human beings were unnecessary for the proper order of things. I don't think that's what's going to happen myself. For one thing, as far as I can see, machines are unmotivated even if they are super-intelligent. Their motivations have to be provided from the outside, presumably by us, and, therefore, even if after we made the super-intelligent machines, they then start improving themselves or inventing new things, it will be as much our invention as it will be theirs. I think it all depends how you look at it.

CARAS: Do you think that the super-intelligent machine will contribute to man's ability to re-order his biology?

FEINBERG: Very probably. Intelligent machines would be helpful in the business of genetically improving human beings, unless we figure out how to do it before the super-intelligent machines are invented. I think it's an interesting race involved as to whether super-intelligent computers or supermen are evolved first. I am not sure which side I would bet on, but I think that both of them are pretty good bets. I think the super-intelligent machines might be helpful in connection with extraterrestrials. Even if it is not possible for us to understand the extraterrestrials or them us, it is, I think, more likely that our machines can understand their machines and perhaps explain to their masters what it is that each one is doing.

CARAS: Can we rely on machines to be always without motivation and if they are motivated could we rely on them to help us re-order our biology?

FEINBERG: You mean the machines might think that they wouldn't want to help produce supermen because they would then be rivals to the machines? That sort of inverse twist to the usual way of thinking? Well, the question of machine motivation is a very complicated one and I think that it's hard to answer it until we're somewhat further along the line of producing an intelligent machine. The developments which I had in mind that we know of right now which might help to make machines that are at least in some ways as intelligent or more intelligent than human beings would seem to me not to have any correlation of motivation. That is, right now, I can imagine developing the machines we have now so that they became much more intelligent but still didn't have a motivation. Whether motivation would come in along the way somewhere is hard for me to say. I prefer not to speculate about it.

CARAS: Is there anything else you want to add, Doctor?

FEINBERG: Well, I think one problem which is connected both with intelligent machines and with biologically improving human beings is – do we want to do it? Even if it becomes technically possible to make intelligent machines or supermen we still have to decide what purpose it will serve. The purposes until there are these machines or supermen around are still those of human beings, and I can imagine the human race might decide that it doesn't want to make these creatures. I myself think that would be a bad idea but I don't believe this is something which any individual scientist, least of all myself, has the right to decide. I think decisions like that can only be made by the human race as a whole and I think it is by no means too early for us to begin worrying about the question of how we decide things like this. There is a whole class of things that I call world-shaking decisions – like whether super-intelligent machines should be made – or if we receive signals from some extraterrestrial life form should we answer them? I don't think any of these questions can be answered without a better idea than we have now of what the aims of the human race are and I think for that reason it would be a very good idea if people everywhere devoted some of their time to thinking along those lines.

CARAS: Can you think of any major decision of this type that has ever been made by the human race? Has mankind ever decided not to do something once he found he could?

FEINBERG: I don't think that until now mankind ever has refused to do something really important that was technically possible for it. There have been some irreversible changes in man's way of life in the past, such as the introduction of agriculture eight thousand years ago or the introduction of industrialisation about two hundred years ago. They just came about in the natural progression of technology and science. I think that the difference now is that we can foresee, if only by twenty or thirty years, the fact that some of these decisions are going to have to be made. And I think as long as we can foresee this we should at least think about whether we want to do it.

ROLL THREE / TAKE TWO

CARAS: Do you really think that the human race is capable of denying itself, or denying its curiosity, and not moving forward with the technological potential once it opens up? Do you really think the human race is capable of ignoring an artificially produced extraterrestrial radio signal once it was identified as such?

FEINBERG: Well, I would hope that there was nothing inevitable about anything the human race does, particularly in connection with something that might have such world-shaking consequences as establishing communication with other stellar systems. The point, I think, is that it is not possible for us to do everything that is – doing one thing by itself precludes the possibility of not having done that – or doing other things which the consequences of the first action leads you to. What I am trying to say is that when we take an action which really can have very important – perhaps irreversible – consequences it would be worthwhile for us to think through what some of those consequences might be before going ahead with it. Now, even after doing that it is possible that we wouldn't be able to keep ourselves from doing the thing – even if we decided it wasn't a good idea – but I would hate it if we hadn't at least gone through the process of rational thought to decide what might come out of it.

CARAS: Do you see anything wrong – any negative factor in communicating with extraterrestrials at this point if they exist and if a means to communicate with them were found?

FEINBERG: I can imagine possible consequences of communicating with extraterrestrials which might be bad. I myself don't believe it very likely that any of them would come to pass. I think on balance that the effects on human beings would probably be all to the good but it is not something that I believe one can be too dogmatic about and for that reason I would like it if a lot of people thought about it before we did anything. Luckily, communication with extraterrestrials is not something that has to be a rush job – since if you use radio signals it takes many years to signal back and forth, so I don't think it would be terrible if we, say, picked up signals from another stellar system, if we spent five or ten years thinking about what the consequences would be of answering them back before we did it.

The 'small' hand-held Panavision camera

Constantine D. J. Generales Jr

[International Consultant in Space Medicine Affairs; Chairman, Section on Space Medicine of the Medical Society of the State of New York]

SCENE ONE / TAKE ONE

CARAS: Doctor, twenty-five years ago, if a graduate student in astronomy told his professor that he wanted to discuss in a paper extraterrestrial life he would have been told to leave the room. This is true. Today, scientists in general believe in extraterrestrial life and are studying it and expect to make contact. What knowledge do you have? What feelings do you have on this subject?

GENERALES: Well, my personal feelings in the matter of extraterrestrial life as a whole picture seem to coincide with the majority opinion. There is a possibility that extraterrestrial life exists beyond our Earth. However, I feel that life if you apply it to man or in the shape and form of man, is unique and characteristic of Earthly beings. I think, however, that as we get to explore the universe we may run into a surprise that there are different kinds of beings who may be closely related to us, but not in the shape of man as a biped and with the intelligence he has.

CARAS: If the morphology was totally different, if in looks they were totally different, could they not be as intelligent or, if they are on a much older planet, could they not be much more intelligent than we are?

GENERALES: That is a great possibility. We would first have to determine when we speak of intelligence just what we mean by intelligence. It is a hard thing to differentiate because when we speak of intelligence today, unfortunately, we confuse it with the mechanisations and the instruments man has invented and there are certainly people two thousand years ago who had no recourse to machines and instruments who were equally as intelligent as we are today and if we are to establish, try to establish the fact, that intelligence exists on other planets. If somebody from beyond the Earth was looking at us at the time of Christ there were certainly a lot of intelligent people in those days but there was no way of recording them.

CARAS: Could you envision life arising on any planet, Earth or any place else, and not undergoing evolutionary sequence?

GENERALES: No! I think evolutionary thinking is on a cosmic scale because the old Greeks used to say and still admit today that everything flows. So therefore, everything is subjected to change, be it fast or be it slow, so I certainly believe that man himself will change like everything else is changing, for the better or for the worse.

CARAS: If life arose on a planet other than Earth and did get launched into an evolutionary sequence, was not static, would intelligence of some form be the inevitable result if the time were protracted enough?

GENERALES: The evolution of man is a very gradual procedure. It is dependent on a number of factors, some of which can be indigenous or shall we say within the body itself and others can be the influence of cosmic factors which we know nothing about. I say this with a certain reservation because we are just beginning now to understand the influence of magnetism on the Earth and its influence on man, but if we go far beyond it, there are certain areas like the cosmic year which is 200 billion years, this designates the one revolution our galaxy makes. We know there have been changes five times in the history of the Earth whereby the temperature of the solar system, particularly the Earth, has changed once every 200 billion years which corresponds to one revolution of our galaxy. Therefore, there are certain changes exclusive to mutations and there are other certain changes that man is going to force upon himself. All these will play a big factor in the existence of man on this Earth and in areas beyond it.

CARAS: What are your views, Doctor, on the origin of life on Earth?

GENERALES: The origin of life on Earth is a very difficult thing to say with any amount of exactness because we do not know a thing about it. We have definitely in the past two decades or so demonstrated that we can synthesise, let us say, amino acids from inorganic matter like oxygen, hydrogen and methane, ammonia, nitrogen, and so forth but the idea is, how were these things put into a cell and how was this cell, from one cell, how did it multiply into specific tissues and these tissues into organs? Then again, as far as man is concerned, as far as we know, there are two sexes in man. How did sex come to be in the picture? We have always spoken about the idea of creating life in a cell but we know nothing about the combination of the factors that made man as man, as a male, and made man as a woman, as a female. That is a great mystery.

CARAS: How well equipped as an organism is man to travel in space?

GENERALES: Man is a flexible being. One of his guarantees for his existence is his adaptability to environment. That is the outstanding feature. His only other, shall we say, competitor, is the insect and whether the insects out do man or whether man outdoes the insects remains to be seen. Insects do not live by intelligence. They have come down through the millennia, through adaptation of some sort to the various strains and stresses of the environment. So, we will see whether the insect or man wins as far as life is concerned on the universe.

CARAS: The space age is here, man is going into space, is going to continue going into space. Now, it is expensive and it is difficult because of life support problems. Eventually, will there be surgical alterations to a living man? Will there be genetic engineering to alter man so that the man can exist in space environments without all the tremendous support equipment that is now necessary? Or shall we look forward to forever man being surrounded within a casket, in a coffin?

GENERALES: As I believe that the origin of man does not necessarily have to have its roots on the Earth, I think that the idea of the panspermia theory of Arrhenius whereby seeds from space are settled on the Earth. I frankly believe that it is also possible for man by a combination of factors to exist in areas beyond the Earth. He can adjust himself to a great amount of heat as well as adapt himself to a great amount of cold. He can get along with less oxygen or with more helium, shall we say. There are combinations of factors there that make man quite unique. Now, as far as his organs are concerned, the time is at hand when practically every organ will be synthetically applied to man and the one I believe that he will value the greatest will be that of his brain. All the other organs, in my opinion, are secondary, are support to the brain which is the most valuable thing. The universe and man's brain are the two most precious things that come to my mind at this moment.

SCENE TWO / TAKE ONE

GENERALES: May I quote you a saying from Aristotle? Over twenty-three hundred years ago Aristotle said, 'If human beings could be shaped by their environment they could change themselves in equal measure by their own efforts.' This is the first time, the first instance, where an authority has mentioned the fact that man may be able to explore other regions besides the Earth.

CARAS: What will it be like for very, very long periods of weightlessness? Will this be counteracted in space travel? Let us talk about weightlessness, long-term weightlessness. What happens to a man, what is going to happen, what will be done about it in long deep space probes? What about on the Moon?

GENERALES: The subject of weightlessness is in my opinion the most intriguing of the factors that man will have to deal with. I have been busy with weightlessness for close to three decades and I feel that the issue involved here is of great concern to us all. There isn't a single person on this Earth who hasn't experienced weightlessness, although he may not have been aware of it at any one time or another. Man weighs less in the second floor of his home than he does in the kitchen downstairs and as you go away from the centre of the Earth, the weight of a mass – man's weight in this case – is less and less and as he starts leaving the Earth and orbiting around the Earth in direct or indirect flight to any part of the solar system and beyond he will be in a state of weightlessness. Whereby it is meant that there is no way that you can register the actual weight of a certain subject or man. So much so that a feather or a piano have the same effect as far as weight is concerned. Now then, man has been adjusted or has adjusted himself to a certain amount of weight on this Earth which is, as we say, one G every time he changes, whether he is in an automobile accelerating or whether it is decelerating, coming to a stop. There are certain changes within his body that take place of which he may not be aware of. If these changes in velocity increase, there will be changes also in the production of secretions within the body ranging between cells and cells or tissues and tissues or organs and organs, all interconnected with the so-called vascular system and the vascular system, of course, we know controls the various organs and particularly the muscles. Muscles thrive on exercise besides the nutritional value they get from the blood in the lymphatic supplies. Now then, we know that there is very little effort exerted, particularly if man moves his arm or moves around in a state of weightlessness, and there is a tendency there that even though the circulation of the blood may or may not change in velocity there are factors which we suspect today that may be detrimental to him, and one of those is the production of what we call atrophy which is a loss of substance in the muscle and which in turn again does nothing very beneficial for man and on top of that we will have, let us say, a lot of physical factors that will play a role ranging from cohesion, adhesion, capillarity and a number of other technical physical terms that we have no need to go into at the present time. All these factors will be combined in my opinion to raise problems which will take at least one generation to get the proper answers, and the three big issues are these: Is it best for man in the long run to keep himself at the gravitational force that he has been accustomed to on Earth – of one G or maybe a portion of that – like $1/16$ or $1/3$ or, still better, should he be born in the state of weightlessness? The agreeing factor rises from the issue: How will man grow? Anything that grows in the state of weightlessness has no direction in which it will grow so, therefore, we will have a 360-degree possibility. They can grow in any direction and the problem then arises: after growth can grow in any direction, eventually man will have to land on a planet which, of course, has gravity. There is where the great problem arises in having man readjust himself to that particular gravitational force. If it is a hypogravity planet, by that I mean if the planet has less gravity than the Earth, then it will be easier for him to get re-adjusted. If it is on a planet like Saturn or Jupiter, where the gravity is greater than the Earth, the problem is exceedingly more and more severe. So, therefore, we have crystallised these issues down into what do we do if man after he gets through with the weightlessness state, whether he travels five or ten, fifteen or twenty years, or all his life, depending on – was he already mature when he started going to a weightlessness state when he was not necessarily growing any more or whether he was just a small infant growing and getting bigger and bigger? So, in my opinion, we have to allow at least one generation and that generation can be counted on from the time we decide to put a baby at birth. At that particular moment we will be able to commence having data. What will happen for man from infancy until he gets to be about twenty-one years old? Then we will have more answers than we can give at this moment?

CARAS: What are the psychological problems of long-term space travel, space-probes?

GENERALES: The psychological factors in my opinion are not as severe as was once thought them to have been. I believe that the psychological factors are not different than any of the anxiety factors that man has experienced on Earth. Anxiety is anxiety whether you are isolated in a dark room on Earth or whether you are isolated in a dark room somewhere else. The control of the mind is the most intriguing problem of the whole anatomy, shall we say, and I feel there is a way of selecting astronauts. There is a way of selecting passengers who are going to fly and as I said before, some of us don't know how the mind works on Earth, let alone how it will be in space.

CARAS: What about hibernation? Is that going to be a factor in deep space travel?

GENERALES: Personally, I have never favoured hibernation from the point of view of space travel. What I mean by hibernation is putting a man, Homo sapiens, to sleep by reducing his body temperature to an accepted value where his metabolic rate is greatly reduced and he can still be alive. That takes all the fun away from space travel. This is not the way to make space travel popular. I recall twenty-five years ago, right here in New York at Mount Sinai Hospital, experiments were carried out by using hibernation to cure cancer. We gave that up because there were problems that arose there, and this I recall very vividly. So, having lived with this problem of space flight and how to get to different planets for many years, I would like to approach, to make the environment of man to conform with the habits, with the strains and stresses which he can tolerate as we know them on Earth. Anything beyond that, it is because we have not got the right answers for that particular situation or situations.

CARAS: Then you don't think that hibernation will be the answer to any of these situations?

GENERALES: I definitely do not agree with the concept that hibernation will give us the answers that we want. Experimentally, we can profit a great deal but to use that for space flight it ends upon who is going to put who to sleep for how long and how is he going to be wakened up?

CARAS: I believe this technically probably could be solved like any other.

GENERALES: The technical issues involved in hibernation are very many. I believe that Old Greek proverb, 'Try everything – only good will result.' From that standpoint I agree but not from the point of view of this early aim of making space travel popular. This is not the way to do it. I agree, however, that using frozen spermatozoa, for example, which we know we can reduce to a temperature level of almost absolute zero and keep them there for weeks and months, that that is a way we can propagate life again, injecting into the female and cause birth. I agree that that is a way we could do it. I also agree with the concept that we can create life extra-corporeally. Italian scientists have to this day created life up to [?] years of an embryo outside of a woman's womb, completely outside the body, and I know the Chinese are working on it and it is not inconceivable that woman can be spared the labour of childbirth if we are able to – synthetically – pick out the best, provided we know what the standards for the best are and then bring a male and the female seeds together to create life. We have enough problems on Earth by having, let us say, haphazard crossings of human beings and I think the time has come to be a little bit more selective with the process of breeding.

SCENE THREE / TAKE ONE

CARAS: Doctor, the population problems on Earth. State them slightly and then lead in to how it is going to justify or require the colonisation of the Moon, then the planets, and what life will be like and what people will be like after? What will the genetic variation be?

SCENE THREE / TAKE TWO

GENERALES: There is another factor that I think will determine whether we shall stay on this planet or proceed elsewhere. In the first place the fossil fuels of this Earth will be mostly and completely used up in about four hundred years, four to five hundred years. In this period of time according to the present calculations the human population will be of such a density that it will hardly allow man to raise his hands. It will barely separate him from the shoulders of another man. We will have to review our golf playing, we will have to review the width of our streets, we will have to review what we will do with our cemeteries, we will have to talk about how high man can live and how deep under the surface of the Earth he can live and even when all that is explored satisfactorily man's population growth still will not be checked and no matter if there is just one or two children in the family, so the time will come sooner than we think. With the science of numbers we will determine man's fate and I firmly believe by then that there will be frequent communication between the Earth and other habitable planets and when I say habitable, I do not mean just those planets confined to the solar system. Man will have reached the speed of light and possibly will have exceeded it because there is nothing to say that Einstein's Special and General Theories of Relativity are the final words as far as man's fate is concerned and, looking into the future, so whether we like it or not there is one way whereby man can contain himself, shall we say, with the cells he has. Medical science has thus far been able to create, let's say organs, by simply rubbing together and dividing into very small pieces shall we say of kidney cells, liver cells and muscle cells, and bringing them together under certain conditions. We will find that the same cells get together and seek out their own kind, so it is not inconceivable that we can start off with just cells that didn't get together and form man. This is not a far-fetched dream. This actually has been done and there is reason to believe that along with creating life synthetically, in a tube, a more appropriate way would be, a more simpler way would be, by getting variable cells together, keeping them at the proper temperature and bringing them together and they would be able to form organs which can in turn also be used to transplant the organs. We have also the issue of making synthetic organs. There is nothing wrong in that but the whole essence, as I see it, is to do everything we can to preserve what is to man the most precious thing that he has, and that is his mind, by virtue of the brain. So, therefore, man will have to do some hard thinking, re-evaluating, agonising in many ways, as to what to do with his future and even though many of us will never be around there is no way of knowing that the consciousness of man, yours and mine today, may not be around one day to see what happens a thousand years or so from this moment.

CARAS: I know you touched on this before but I would like to get your personal opinion. Answer me as 'I think' or 'I don't think.' Seriously, do you think, when we get out into the solar system and beyond and we reach planets either by radio or by manned space flight or by unmanned space probes we will find evidence of extraterrestrial life either there now or – you are an archaeologist, obviously by hobby – at least do you think we shall find evidence that life once existed elsewhere than on this Earth? What argument is there against it?

GENERALES: The only argument people can have that there may not be life beyond the Earth is simply their ignorance and as Shakespeare once said, 'The only darkness there is, is that of ignorance.' The logic type mind of many scientists today leads us to a dead end. I feel that the greatest contributions can be made by two kinds of people, the dreamers, which we have had plenty of in the past, and also the specialists who look at things from a very small viewpoint, but very penetrating. The synthesis of knowledge, taking the knowledge from the specialist and applying it to the general field, is where the fruits for the benefit of mankind will come. I am of the opinion that as we go on and on and as more universities are being founded, I have a great faith in the university system, the more children are allowed to explore their ideas and fancies in their early childhood and the more we expose them not only just to science but also to the humanities – and I sincerely hope that one day a language will be found whereby the communications will be simplified so that even the Chinese or any of the Oriental languages can be brought together with the Romance languages so as to save time in communicating ideas and knowledge – then we will be able to proceed much faster. I think that in the extraterrestrial sphere we can make no limitations. The universe is boundless and, as I said previously, it is my feeling that man's future lies beyond the Earth. There is no reason to believe that if this solar system should come to an end, which it may very well in four to five billion years, that man's existence has to come to a stop. If we believe in a beginning of life, as some people do, there may be also a cessation of life but I believe in a cyclic sort of a theory where everything goes round and round and round. Nobody knows where we are going. I think that most of us believe that wherever we are going, we are getting there very, very fast.

CARAS: Do you want to make some kind of a statement about extraterrestrial life or space travel?

GENERALES: I firmly believe that the time is close at hand when the ordinary propulsion of chemical methods will be one of the past. I look forward to an electromagnetic propulsion along with atomic propulsion which will enable man to leave the Earth in a horizontal fashion with proper acceleration, noiselessly, without any air pollution, to reach the planets and beyond in very large spaceships that will contain between fifty and a hundred thousand individuals. They will be self-sustaining and in perpetual flights of this sort where the spaceship is so large that it cannot land on a planet, that there will be smaller spaceships that will be picked out and purposefully be allowed to land on certain habitable planets which can communicate without trouble by taking off from these planets back on to the mother ship, and these spaceships will be electromagnetically shielded against any kind of lethal ionising radiation. These ships will be of such a magnitude that they will be in perpetual flight and will never land on Earth again. They can regenerate themselves. There will be life created on these spaceships in test tubes. There will also be huge atom [?], everything can be constructed and will be self-sustaining. When we reach that state, I believe in the next, maybe, one to two hundred years, man will be able to free himself completely from the necessity of carrying water with him. He will be able to combine the hydrogen and the oxygen atoms or radicals in space and make water. When he does that he has solved a very, very major issue and as medical science goes faster and faster you will find that there will be a tendency to propagate the better and the best things in man so that we will have a society that will have eliminated most of the [?] which causes dissension on this Earth so that when man meets on another planet there will not be any hatred. All the negative emotions will have been left aside and only the ones that we desire for the benefit of one man to another, that these particular emotions and factors and virtues will be able to propagate and as there is very little hope on Earth that this can be accomplished on Earth, I hope that the habitation – I like that word, habitation – better than colonisation – that the habitation by man of himself on other planets will give us a new kind of society where there will be eternal peace.

A posed publicity shot with Arthur C. Clarke on the left and Victor Lyndon on the right. Lyndon was the associate producer at the beginning of *2001*'s production. The office is actually Lyndon's. SK thought his sprawling and cluttered office might give the wrong impression.

Irving John Good

[Trinity College, Oxford; Atlas Computer Laboratory, Chilton]

22 MARCH 1966

CARAS: What do you think of the probability of contacting extraterrestrial life in the near future?

GOOD: Well, it is possible that we have already contacted it in a sense by seeing flying saucers, although I am rather sceptical about the stories myself, they are very difficult to evaluate. However, I think the chance of contacting them in the near future is greatly increased by the interesting condition of our own civilisation at the present time. We are just about to take off into space and, I believe, to invent the ultra-intelligent machine and, when we have invented a machine which can itself design even better machines, there will certainly be an intelligence explosion and our civilisation will then advance at an enormous rate so this means that we are, at the present time, in an extremely interesting condition which must occur only once in the history of civilisations which live for hundreds of millions of years.

CARAS: Do you personally believe in the existence of extraterrestrial life? What conceivable forms might they take?

GOOD: I believe in the existence of extraterrestrial life very strongly. For one thing, there are a 100 trillion stars in our own galaxy and the Sun is merely a star of average age, so there will be a very high proportion of these stars which are older than the Sun and if they have planetary systems [1],* which seems probable, they will have, in a certain proportion of cases, the conditions required for life and since they are much older than the Sun, the civilisations that will have developed on these planets [2] will be hundreds of millions of years ahead of our own so that the life that will have developed will be not merely intelligent but vastly more intelligent than our own. It is very difficult to imagine what shape the beings or entities (as perhaps they should be called)

---

* Figures within brackets in the text refer to Professor Good's notes that follow on immediately after the interview.

would take. They might be machines, they might be crosses between machines and people and even, not to be dogmatic, they might be disembodied spirits.

CARAS: What do you think our own civilisations would be like if we were a million and ten thousand years old, instead of a ten-thousand-year-old civilisation?

GOOD: I think it will be always as difficult to imagine what our future would be so far ahead in time as to imagine the life in other parts of the galaxy.

CARAS: You mentioned before that there might be a man or might be part-man machines. At what point does man cease to be a man and a machine start to be a machine? Is there a clear dividing line between the biological and non-biological entity?

GOOD: As far as I can see, it is a question mainly of whether the machine, if we are going to call it a machine, contains a great amount of organic chemical material in its structure. I feel that a cogwheel machine would not be regarded as a living organism however complicated it was and however much it behaved like a living organism. I think we would tend to balk at the notion that a large cogwheel machine could be conscious and feel pain.

CARAS: Do you think there is anything in this division between man and machine that upsets the long-standing theories of God and the universe and so forth?

GOOD: I think there are certainly a number of interesting speculations that one can make about the nature of God or rather the definition of the word God. I believe the ultra-intelligent beings that permeate the universe might very well be in telepathic communication with one another and that they might even exist as an integrated, more or less immortal consciousness permeating the whole of the universe. If this is so, this could be regarded with a slight stretch of the word, as a definition of God [3]. This network of communication would resemble the network of ourselves in our own brain and would have the same potentiality of integration provided that the communication could be made fast enough. This is a serious difficulty as a matter of fact because the velocity of light, although it seems fast on Earth, is not all that fast by galactic standards. It takes about eighty thousand years for light to cross the galaxy [4] and problems of communication would be very serious unless signals can travel faster than light.

CARAS: What do you think that the possibilities are that ultra-intelligent entities have achieved faster than light capabilities?

GOOD: According to the Special Theory of Relativity, no signal can travel faster than light unless signals can also travel backwards in time. Now there are, oddly enough, some indications even in physics that signals can travel backwards in time and also in some of the best controlled experiments of telepathy, the telepathy was precognitive, that is to say it did go backwards, so that there are indications that the Special Theory of Relativity is slightly misleading, I should say [5]. That is to say that some people interpret it to mean that no signal can go faster than light whereas the correct interpretation is that if signals can go faster than light, then they can also go backwards in time or can go with infinite speed. I think the best hope for this integrated immortal consciousness to exist is if telepathy has been developed to such an extent that communication throughout the universe can be instantaneous. This, unfortunately, leads to certain time paradoxes, but I believe these can be resolved [6].

CARAS: Well, we have covered the attainment of speed of light, telepathy, extraterrestrials and so forth. What do you think is the most startling revelation among these or any others that you wish to bring in that we are likely to encounter in the next century?

GOOD: It is difficult to think of the most startling because what in fact is liable to happen will, itself, possibly be beyond my imagination so I really have to let my hair down to think of what would be the most startling events that could occur. I suppose I could try to contemplate science fiction ideas, but I think they would be more or less of a standard type. As I said before, there might be beings that are disembodied spirits and we might find ourselves receiving signals from apparently nowhere, giving us information about the universe. One interesting question of course is why it is that these ultra-intelligent entities have not previously contacted us in a convincing manner? It does seem to me that they must have some reason for not wishing to allow us to enter the cosmic club at the present time, perhaps because we are backward ethically, perhaps because we have not yet reached a sufficient level of technology to be worth entertaining as members.

CARAS: If some direct evidence of extraterrestrial intelligence suddenly presented itself on earth, what do you think the psychological effects might be on the world in general?

GOOD: I think it is possible there might be a certain amount of panic or there might be something in the nature of a religious revival.

CARAS: But do you think there would be any danger of cultural shock, as has been referred to by other people?

GOOD: There is a precedent for this notion in Fred Hoyle's *The Black Cloud* when an attempt is made to educate a scientist by the Black Cloud and the effect is that the ideas are so novel to him that he has to sacrifice all his previous understanding and, in fact, goes mad, so I think that possibly there is a danger of a social shock. All our standards of social living, rather than our scientific ideas perhaps, our standards of ordinary life would be undermined somewhat.[1]

CARAS: Do you think the governments would possibly withhold any early inkling of proof of extraterrestrial existence? I'm not saying that if a flying saucer landed in Central Park, but if there was some information they had that they could withhold for a while, do you think they would be very careful about announcing it?

GOOD: I think they would withhold it for some time until they had thought out the consequences of releasing the information. Of course, the United States Air Force has recently given a fairly large sum of money to Colorado University. I believe it is to research into flying saucers and it may be that they've done this in order that they should not be accused of withholding information about them.[2]

CARAS: If the ultra-intelligent beings in the universe have such unimaginable power, might not contact with them prove to have some dangers as well as the hoped-for utopian gains that one talks about?

GOOD: I think that, ultimately, when we are ready for it, contact with the galactic community will be only beneficial, although there might be some shock at first, but this depends very much on whether the galaxy is really under proper governmental control, whether there is a cosmic club that is concerned with the advantages of its members, rather than a free-for-all, and I think that if we merely contact life in outer space and discover that the galaxy is not under proper control, that there would be very serious dangers of war

---

1 Fred Hoyle (1915-2001), the Cambridge astronomer, wrote several novels. *The Black Cloud* was first published in London by Heinemann in 1957. It has oft been reprinted. Hoyle developed the steady state cosmological model with Hermann Bondi and Thomas Gold in 1948 and, interestingly, coined the term 'big bang' as a dismissive description of the rival theory, a description that has remained and shed its dismissiveness [7].

2 This was the Committee for the Scientific Investigation of Unidentified Flying Objects led by Dr. Edward U. Condon at the University of Colorado. The university was appointed to investigate the subject after the House of Representatives pressured the USAF for an independent study of the phenomena. Their findings were published: Dr. Edward U. Condon, *The Scientific Study of Unidentified Flying Objects* (New York: E. P. Dutton, 1969), popularly known as the 'Condon Report.' Condon had been handed a poisoned chalice and subsequently got it in the neck alike from those who believed and those who disbelieved in flying saucers. SK in his September 1968 *Playboy* interview described the *Report* as 'neither serious nor searching.'

between ourselves and other civilisations. I don't, in fact, believe that this is the situation because if there were criminal civilisations abounding in space, I believe we would have been attacked long ago.

CARAS: Isn't it possible that they just haven't got around to attacking us?

GOOD: It is possible, but the galaxy is very ancient, several thousand million years old and our own civilisation is only a few thousand years old. Men have existed for only two million years. I would have thought that in the course of the last few million years we would have been completely wiped out if there was a gangster civilisation loose in our vicinity.

CARAS: Well, if we assume that there is this galactic community who have organised the galaxy, wouldn't it be reasonable to assume that at some time in the Earth's past, it has been visited by extraterrestrial explorers?

GOOD: I think it is quite possible, in fact, probable, that we have already been visited by explorers from the galaxy, but they have not left, apparently, very much trace unless the human race itself is the trace in question.[3]

CARAS: Do you think these extraterrestrial explorers may have left some monitoring device in the solar system somewhere?

GOOD: I should think it is extremely likely that some monitoring device has been left in orbit around this Earth, probably several of them and probably staffed by ultra-intelligent machinery or even some of the top entities themselves. It is very easy to overlook the fact that a very advanced civilisation, hundreds of millions of years ahead of our own, would have perfected medical science and have achieved personal immortality, barring accidents, so that it wouldn't really be a very great hardship for these top entities to live for even millions of years in orbit round the Earth. It would still be only a small fraction of their entire lifetime. Of course, when I refer to these monitoring devices, I'm not necessarily referring to flying saucers because I feel that the evidence for flying saucers is rather difficult to evaluate and that they may or may not exist. I feel that one should suspend judgement for the time being on that question.

CARAS: One has heard the view expressed that enquiries along the lines of extraterrestrial intelligence and other forms of extreme scientific speculation are profitless and virtually of no use whatsoever. Do you agree with this?

---

3 This, of course, is Charles Fort's idea that 'I think we're property.' A theme taken up by many science fiction writers. It has been suggested that our DNA could be the extraterrestrial artefact we have been looking for all these years.

GOOD: I think that speculation is stimulating and interesting and I think that people who insist on only hard facts in discussion are merely cutting out one of the more interesting philosophical aspects of human thought. There are other reasons for being interested in speculation about extraterrestrial life at the moment and that is that we ought to be preparing ourselves for the possibility of meeting it.

CARAS: What is the present view as to the actual physical size of the visible universe?

GOOD: It's something of the order of say 10,000 million light years, which incidentally raises a rather tricky theological question. Namely, if one thinks of God as being a part of the universe, or the whole of the universe, rather than lying outside the universe, how does He succeed in communicating with Himself? This, of course, is only a problem if we assume that no signals can travel faster than light. There is a certain amount of research going on at the moment on what is called machine intelligence or artificial intelligence. For example, there has been a very effective checkers playing program written in New York by Dr Samuel,[4] which plays a reasonably good master game, not world championship standard as yet, and there are other indications that machines will gradually take over intellectual activities which, in the past, have been regarded as typically human. In fact, Lord Bowden once on television said that it might be possible to construct a machine with the intelligence of a human being, but it might be more economical to make a brain in the usual way. However, this overlooks that at double the cost you could make a much bigger machine that would perhaps be ultra-intelligent. By an ultra-intelligent machine, I mean a machine that can perform every intellectual activity better than any man. In particular, of course, it could design other even more intelligent machines, so that as soon as we have the first ultra-intelligent machine there will clearly be an intelligence explosion and the developments in a hundred years following that will be utterly fantastic.

CARAS: You are using the word ultra-intelligent machine to be synonymous with what we call computers today, but do you think that computers will approach human personality traits as they begin to learn our traits?

GOOD: I think computers might very well develop personality traits, good and

4 Dr. A. L. Samuel, 'Some Studies in Machine Learning Using the Game of Checkers,' in *Automated Programming*, vol. 6, part 1 (Oxford: Pergamon Press, 1969). The paper was first published in New York in 1963.

bad, because there are indications from people who have studied organisations, that when they become very large they begin to develop faults that are very difficult to track down, so the notion of a neurotic machine is certainly not, by any means, unreasonable [8].

TUESDAY 22 MARCH 1966 / TAKE ONE

CARAS: Doctor Good, do you personally believe in the likelihood of our contacting extraterrestrial life in the near future?

GOOD: I believe extraterrestrial life will contact us when they think the time is appropriate, and I believe that this will happen fairly soon, not merely because of space exploration, I think it's because…

TAKE TWO

CARAS: Doctor Good, what do you think of the probability of our contacting extraterrestrial life in the near future?

GOOD: I believe we will contact extraterrestrial life in the near future, but rather that it would perhaps be better to say that they will contact us when they feel the time is appropriate. I believe that will be fairly soon, partly because of space exploration, but perhaps more because of the advances in computers, the advent of what one could call the ultra-intelligent machine. This, I think, will occur within the next thirty years or so.

CARAS: What form do you think this extraterrestrial life might take?

GOOD: It's almost impossible to conceive because the life that must exist in outer space, much of it will be hundreds of millions of years ahead of us in development and for all we know they might even be disembodied spirits.

CARAS: What do you think our own civilisation would be like if we were a million and ten thousand years old instead of ten thousand years old?

GOOD: In a million years from now, it's almost impossible to conceive what men will have become. We will have changed. Our descendants would perhaps not be called men because they would be a different species. I believe they may have developed telepathy so that they would be in constant communication with one another. This would mean that all our descendants would act as a kind of integrated consciousness and would not be separate individuals. Furthermore, I think that there would quite likely be a telepathic communication with all the other members of the cosmic club. That is to say, all the other ultra-intelligent entities in the universe, or at any rate in our own galaxy.

CARAS: At what point does man cease to be man and a machine starts to be a machine? Is there a dividing line?

GOOD: It's difficult to find a dividing line between men and machines. Suppose we construct chemical objects which at first we would call machines. If they appeared to have intelligence and consciousness as Satusi Watanabe once said, we would probably call them man-made animals rather than machines.

CARAS: What does this do to the long-standing theory of God or gods in the universe?

GOOD: I think the ultra-intelligent beings, in telepathic communication with one another might almost be taken as a definition of God, although of course it would not be consistent with other definitions.[5] This idea of God, this concept, would be a god that at any rate had not created the universe although it would be incomparably more powerful than anything we can imagine [9].

CARAS: What do you think is the most startling revelation that we are likely to encounter in the next century? Do you think it will be this contact with…?

GOOD: Yes, I think that the contact will be made within the next century and it will certainly be startling. It's difficult to think of anything that could be more startling than that. There will be great fear, perhaps, but I imagine that this God, as I was calling it, would somehow be able to allay our fears by a full understanding of human psychology.

CARAS: [?].

GOOD: Our galaxy alone has a 100,000 million stars in it and there are about ten thousand million galaxies in the observable universe. Moreover, it's considered nowadays that many stars resemble our own Sun in having planetary systems. I would therefore assume that there are certainly trillions of civilisations, although it is possible that all the ultra-intelligent civilisations in our own galaxy have descended from a single civilisation, which was ahead of the rest.[6]

---

5 These remarks of Good's had a profound influence on SK and he was to echo this very thought when being interviewed: 'I will say that the God concept is at the very heart of *2001* – but not any traditional, anthropomorphic image of God. I do believe that one can construct an intriguing *scientific* definition of God.' SK, *Playboy* interview, September 1968.

6 Good is here resurrecting a discredited anthropological theory – diffusionism – and on a cosmic scale, yet. See Footnote 1 in the Margaret Mead interview, *infra*.

CARAS: Do you think that life originates individually in each of these situations or do you think it comes from one source? The original galactic dust as it were?

GOOD: Theories of the origin of life are extremely speculative but I believe that no biologist today would presume or appear to believe that life is something which originates naturally, that it is almost inconsistent with the Second Law of Thermodynamics. There ought to be something called the Fourth Law of Thermodynamics which is that organisation increases in little local isolated places [10].

CARAS: With all these advanced civilisations in the universe, isn't it likely that we've been visited before?

GOOD: I believe that we probably have been visited before but at the moment we must be under very constant surveillance because we are in an extremely critical stage, a very interesting stage from the point of view of the top entities. We are just about to join the cosmic club ourselves or perhaps to be annihilated, depending on whether we behave ourselves.

CARAS: Which do you think is the likelihood we have with us for the next fifty years, acceptance by the cosmic brotherhood or annihilation?

GOOD: I believe they will be able to accept us because their control is so great that we would be foolish to do battle with them. I think there must already be a galactic police force, in fact if there had not been a galactic police force we would probably have been annihilated by galactic criminals long ago.

CARAS: Do you think they'll be warning us before we do something foolish?

GOOD: Yes, I think they will warn us, they will make their presence known. They will do it I think in a gentle way, rather than by terrorising us, but they might have to produce some extremely dramatic incidents in order that we should believe what we see with our eyes. People, as a whole, are not going to believe that flying saucers exist. Many scientists will not believe that flying saucers exist even if they do exist. Now, I don't know whether they do, all I'm saying is that it is very hard to break what we might call biological habits of thinking.

CARAS: Where do you think machine intelligence is heading on this planet? What are the next steps? What are the next things we'll see?

GOOD: At first, I believe, computers will do more and more of the intellectual pursuits that we regard as peculiarly human. So far, perhaps the most dramatic piece of machine intelligence is the chequers or drafts-playing program that was written by Samuel in America. This program is capable of beating very good chequer players. I think that one of the next advances will be in chess. It is sometimes thought that chess is easy to program for computers.

This is not at all the case. It will probably be about another ten years before computers can play a really good game of chess [11]. At this stage I think governments will feel that it is necessary to spend hundreds of millions of pounds in developing ultra-intelligent computers because they will realise that this is the easiest way to achieve scientific advances and, even, sociological advances.

CARAS: Would you say ultra-intelligent machine [?].

GOOD: By an ultra-intelligent machine, I mean a machine that is capable of performing every intellectual activity somewhat better than any man.

CARAS: What about neurosis?

GOOD: There is a possibility that the machines will have neurosis and will need to be analysed but since they will be so much more intelligent than men they ought to be able to work out how to do it themselves. Another rather similar point is that the ultra-intelligent machine will undoubtedly create many social problems but, once again, being ultra-intelligent I believe these machines will be able to devise an adequate solution.

TWO / TAKE ONE

CARAS: What evidence is there that there are other civilisations like our own?

GOOD: There is quite a lot of evidence by spectroscopic examination of the stars but one piece of evidence which is often overlooked is that the distances of the planets in our solar system from the Sun obey a rather simple law known as Bode's Law[7] and the near simplicity of this law, the mere fact that there exists such a simple law suggests that the solar system originated not in an accidental way but in a natural way. Even though theories of the solar system are controversial, this fact by itself suggests that there are many other solar systems in the universe resembling our own, even with respect to the distances and the sizes of the planets.

CARAS: Going back to that first question, what are your feelings about the [?]?

———

7 More commonly known as the Titius-Bode Law and sometimes as the Bode-Titius Rule (take your pick). It roughly predicts the spacing of the planets in our solar system. It was first mooted by Johann Titius in 1766 and was expressed mathematically in 1778 by J. E. Bode. It relates the mean distance of the planets from the Sun based upon a mathematical progression. As Willy Ley one wrote, 'Professor Bode could not explain why the rule worked …'

GOOD: I think it's not quite so much a matter of us encountering the ultra-intelligent beings in outer space as that they will encounter us, but they will decide when the time is ripe to let us know of their existence and this, I think, will happen fairly soon, within the next few decades, not so much because of our space travel, although this I suppose is one of the things that will draw us to their attention but more, I believe, because of the advances in computer science. I believe that ultra-intelligent computers will be constructed within about the next thirty years and these will lead to such an enormously rapid advance in scientific explosion and an intelligent explosion that it will be impossible for the top entities to ignore us. They will be forced to let us know of their existence and let us know of their immense power in order that we should not be too obstreperous.

CARAS: Talking about the mind, this dividing line between man and machine. Will man come to this by evolution or will he come to it by building machines that will take over?

GOOD: I believe the intelligence explosion will occur first via computers. It could occur also by a kind of breeding but this would clearly take several generations, well over a hundred years, whereas I believe the computers will reach the ultra-intelligence stage within a few decades. Now this of course will create a very dangerous situation. There will be a danger that the computers will take over, the computers and the machines which they would be capable of constructing. On the other hand, if we can somehow train them so that they will serve human beings – and after all this is quite possible since there's no particular reason to believe that they will have any other ambitions – then we could ask the computers themselves how to solve the social problems that they have created by their existence.

CARAS: How close are we to the point when machines will be counselling us, not answering questions as a computer here in this room does but actually counselling us?

GOOD: I think the moment that machines will become, as it were, an oracle, that they will decide so to speak to give us advice will be within a few years of the construction of the first ultra-intelligent machine. As soon as the first ultra-intelligent machine is constructed, it will be capable of building even more intelligent machinery and this is why I refer to the intelligence explosion. This will happen so quickly and so suddenly that it can hardly be more than a few years from the construction of the first machine to the existence of machines that are capable of behaving as oracles.

CARAS: How will these super machines be?

GOOD: The first ultra-intelligent machine will need to be educated by human beings. Before long, of course, it would be in a position to educate its own teacher but after the first machine there is no obvious reason why men should be required to do the education. The machines would be able to educate one another.

CARAS: Do you think we will have a major scientific breakthrough from an extraterrestrial civilisation?

GOOD: I don't expect them to send guidance to us as to how to construct the ultra-intelligent machine because until we have constructed this machine we will not be any kind of a threat to them and when we have constructed it, then we will be a potential threat because of our aggressive tendencies and it will then be necessary for them to get in touch with us and to tell us the facts of the galaxy and of the universe so that we will behave ourselves in a reasonable manner.

CARAS: [?].

GOOD: I believe they are leaving us alone at the moment for various reasons. One is because we are part of the galactic zoo. They are interested in us as a form of life and we are in an exceptionally interesting state at the moment because we are within a few decades of being able to join the cosmic club.

CARAS: Would you like to speculate again what life will be like in, say, a thousand years from now?

GOOD: The question of what life might be like a thousand years from now is of course interesting and almost impossible to answer. Within a thousand years evolution could not have taken place to any very great extent [12]. Certainly our intelligence and our stature and various other qualities could be greatly improved because a thousand years is forty generations, so I think that in a thousand years from now our descendants will be unmistakably like ourselves but a lot better. They will be men. In a million years from now on the other hand I believe that the word 'men' would not be appropriate as a description of the species that would be descended from us.

CARAS: What effect do you think it will have on our society exposed to that contact? Do you think there's a chance of a major cultural change?

GOOD: There's no question that when we become convinced that ultra-intelligent beings exist in outer space, this will undoubtedly be an enormous emotional shock for human beings and I believe that this shock will be so great that it will be necessary for these beings, these top entities, to avoid this shock by taking into account human psychology, somehow breaking the news to us rather gently.

CARAS: [?] time going backwards. How close are we to this civilisation?

GOOD: Well, there are a number of difficulties for the top entities. They have difficulties of communication, or presumably do, because the diameter of the galaxy is eighty thousand light years. Now, unless signals can travel faster than light, this creates a very serious communication difficulty for the top entities. However, there is a possibility that telepathic signals can travel faster than light. In fact, some of the best evidence for telepathy was for precognitive telepathy. Now precognition means signals travelling backwards in time. This, as it happens, according to the Special Theory of Relativity is equivalent to the existence of signals that travel faster than light. I believe that my speculations about the existence of an integrated immortal consciousness which exists throughout the universe depends very much on the possibility of telepathy and also on the possibility of instantaneous telepathy, of telepathy travelling faster than light, and this does seem to involve us in the possibility of signals going backwards in time [13]. There are some perfectly respectable theories. There are some well-known physicists who have speculated on the possibility of backwards time. It may be…

THREE / TAKE ONE

CARAS: How close are we to a realisation of backward medium time? What kind of effect do you think it's going to have on humanity?

GOOD: One of the physicists who suggested a theory involving backwards time was Feynman, who suggested that a positron is an electron moving backwards in time. It's interesting to speculate on the possibility of whole universes, not in our galaxy but in space, which are as it were negative to ours with regard to charge and so on and which have time going in the opposite direction. It would be very difficult to communicate with such a universe, or perhaps I should say with such an anti-galaxy, partly because of the immense distances, but if we could communicate with such a galaxy then some interesting possibilities of finding out what would happen in the future would emerge.[8] If they could communicate with our future and we could

---

8 It was Paul Dirac who first proposed the existence of a positron in 1928. The mathematical description he provided suggested that it was an electron moving backwards in time.

Richard Feynman has proposed (puckishly?) that there is only one positron in the whole universe and it's moving backwards and forwards in time and giving us the illusion that there are countless numbers of them.

communicate with their present this would mean that we would be able to obtain information about our future.

CARAS: Is the old fictional creation, the time machine, really something that might come about?

GOOD: The trouble with the idea of a time machine is that it seems to be two logical contradictions because it would seem to involve the possibility of your being able to go backwards in time, choosing the past and then coming forward and finding that things were no longer the same as they were. This, as a matter of fact, is not a complete logical contradiction because time might have a peculiar branching structure and we might be quite wrong in imagining that time has a linear structure like a straight line. It might even be possible to change the present by going backwards in time.

CARAS: Do you think it is possible that some of the ancient civilisations on Earth knew more than we know of galactic [?]? About the fact they were possibly visited and know more than we know?

GOOD: It is conceivable that some ancient civilisations were visited by the top entities and it is also conceivable that some of the religious figures of the past were connected with the top entities, but I believe it is most unlikely that any earlier civilisation had developed science to the extent that we have done because if they had done so there would surely have been some trace left of their work.

CARAS: I don't mean knew more than we know by their own development but I mean by revelation.

GOOD: Yes, they might have known more about the top entities, as I say by revelation, but they would not have called it science as it were, they would have called it religion.

CARAS: Do you think it's possible that any top entities are among us now in a disguised form? Is this a possibility?

GOOD: It's quite conceivable that there are top entities amongst us at the moment. There is simply no limit to what they might be able to do. We already know very much more than we knew ten years ago concerning the structure of the genetic substance DNA, as it is called, and it is quite possible that even without the advent of the ultra-intelligent machine, that within a century or so we might be able to construct DNA according to our own specification, consequently a civilisation that was hundreds of millions of years ahead of ours might find it quite easy to produce DNA of great length and produce therefore biological structures to almost any specification. The DNA would be regarded as a kind of computer tape. It would be a programming tape for producing biological entities.

CARAS: Do you see biological immortalities in the future of mankind?

GOOD: I believe that the ultra-intelligent machines would in quite a short time be able to make very serious contributions to the problem of longevity and it's quite conceivable that our own children will live for hundreds of years as a consequence of this or even for millions of years.

CARAS: How will society decide who among them, which men shall live and which men shall die?

GOOD: The problem of which people to keep alive is already present with us even without reference to these speculative notions concerning virtual immortality. For example, we know that there are many people in the world now who are starving and it is quite possible for anyone to take charge of a few individuals and keep them alive, so we are already in this position and also in the medical sphere we are in this position because for example it is possible to keep people alive who have circulatory trouble and yet many people do in fact die of heart failure.

CARAS: Do you suppose ultimately when there is a pill or an injection or a device that can keep people alive indefinitely this will be a decision between the machine [?]?

GOOD: The decision of which people to be kept alive is an ethical problem which is so extreme that I hardly dare say anything about it. I can only hope that the ultra-intelligent machines will be able to solve that problem amongst all the other problems they will be able to tackle.

CARAS: Can you envision the future of man on this planet with ultra-intelligent machines or does the future of man [?]?

GOOD: There might be some kind of a cross between a machine and a man in the future, a sort of bio-machine, and on the other hand the machines might take over, or they might be kept quite separate from humans, I simply can't guess which of these is most likely to happen.

CARAS: What I meant really was the computer, for instance. If development of the computer could stop at this point and we could come to an impact with the computer would this mean that the development of mankind would indirectly be affected or would there be other possible happenings?

GOOD: If it turns out that owing to some kind of engineering problem which is insoluble that computers of the kind I've been describing cannot be constructed, for example if it would be too expensive, then the future of man will be perhaps most easily advanced by means of breeding and for all we know some of the organisations that take care of artificial insemination in some countries are already, without even their own governments' consent, carrying out breeding experiments [14].

CARAS: Will some day the genetic tools, that is biological reproduction as we know it now, be suggested by computers in terms of evaluating a respective mate?

GOOD: Yes. I think an ultra-intelligent machine amongst its many other capacities could act as a matrimonial agency. There have been many civilisations in which people married someone selected other than by themselves, by their parents for example, and I have no evidence as to whether they were more or less happy than people who make their own mistakes.

SK and Arthur C. Clarke showing 'visiting fireman' from MGM head office around the studio

FOOTNOTES TO THE INTERVIEWS:

Irving John Good, 2004

1 It is now known with practical certainty that there are 'close' stars, other than the Sun, having planetary systems.

2 I had written earlier only that it is probable that life will have arisen in some reasonable fraction of these planets.

3 I have said that 'Godd' would be a better name than 'God' for the sense defined as the ultra-intelligent entities in telepathic communication, each acting as a cell in an 'integrated immortal consciousness' corresponding to a brain the size of the universe. The use of the spelling Godd helps to make a distinction between the more familiar supernatural meanings of the word and the more naturalistic meaning. I doubt whether Godd would want worship, but obedience would be wise.

4 Some people take an estimate of 100,000 light-years instead of 80,000, but this is unimportant because the edges of the galaxy are fuzzy. One wouldn't want to enlarge the estimate merely to include relatively few 'outliers.'

5 I am referring here to the very famous experiments done by S. G. Soal (of Queen Mary's College, London). Unfortunately it has been shown fairly conclusively that Soal had cheated, consciously or unconsciously. However, I still have an open mind about the possibility of telepathy and extrasensory perception (ESP).

6 Experiments organised by Aspect in France, related to some theory by J. S. Bell, appear to show that influences can be instantaneous in quantum mechanics. I don't know whether the experiments have been replicated independently.

7 Thomas Gold told me it had taken him about a year to convince Fred Hoyle that the steady-state cosmological model, proposed by Bondi and Gold, was at least plausible. Hoyle gave the theory extremely good publicity.

8 My Windows 98 computer tells lies and often forces me to shut down improperly. The computer doesn't apologise and seems to reprimand me. Such behaviour in a human would be called neurotic. I hope Godd isn't neurotic.

9 A god of only finite power is accepted by H. G. Wells in *God the Invisible King* (1917) and also in one of his novels, *The Sleeper Awakes* (1910). John Stuart Mill, the famous nineteenth century philosopher, had previously proposed the concept of a god 'finite in both knowledge and power' as a solution to the problem of theodicy: this is the problem of why God allows evil. I think more important is, why God allows unnecessary suffering.

10 I no longer believe that the origin of life is inconsistent with the Second Law of Ther-

modynamics. In this paragraph I appear to be talking as a creationist, the creator being Godd rather than God (should the spelling therefore be creationist?).

The little isolated places would have to be very small indeed for entropy to go in the wrong direction in a closed physical system, at least in our part of the universe.

The literature on the origin of life is, of course, extensive. A theory would be well supported if and when uncontroversial life is produced in a laboratory. I guess this will be done before the year 2050.

11 As is well known there is now an IBM chess program capable of beating Kasparov, the human world champion. Writing the program took many man-years. I think it is interesting that C. H. O'D. Alexander, who won the British chess championship twice, had the IQ of a four-star general, told me his belief in about 1950 that no computer would ever play a good game of chess! He didn't know how much more powerful computers would become.

12 In view of the amazing advances in molecular biology, it is quite possible that evolution will advance to a very great extent within the next hundred years. This already creates ethical problems.

13 It is implicit that a necessary condition for the existence of Godd is that telepathy is possible. The 'sufficitude' is appreciable and the 'necessitude' is virtually infinite. For definitions of these two expressions in terms of weight of evidence see my ''Legal Responsibility and Causation' in *Machine Intelligence 15*: *Intelligent Agents* (K. Furukawa, D. Michie, and S. Muggleton, eds; Oxford University Press, 1999). pps. 25-59.

14 Once again man might very well be changed completely by humans or robots operating directly on DNA. In other words, by highly unnatural selection, instead of by natural selection or breeding.

Gerald S. Hawkins

[Professor of Astronomy and Chairman, Department of Astronomy, Boston University; Astronomer, Smithsonian Astrophysical Observatory]

SCENE ONE / TAKE ONE

CARAS: Doctor, I want to ask you about Stonehenge. Could you tell us something about the history of it?

HAWKINS: Stonehenge was built in the second century BC. People began to work there in the date of 2000 BC and I have shown it as an observatory and I have suggested that it is a computer. My observatory theory seems to be unassailable. Whether it was actually used as a computer or not is the centre of a controversy right at this present time.

CARAS: How could Stone Age people have done anything so sophisticated unaided?

HAWKINS: The question has been raised about the mentality of these people. My mind is open. It seems that they did this thing and therefore they must have had some intellect. I believe they had an intellect comparable to our own. Other people have challenged this and feel that the whole thing had been built by chance and that we are stretching our imagination too much to say that our ancestors four thousand years ago were anything like we are today.

CARAS: What about the idea that Earth has been visited in the past and possibly it is earlier peoples' influence? Do you give any credence to this at all?

HAWKINS: I don't believe that we have been visited in the past by any intelligent beings. I don't believe that Stonehenge itself was erected by a master race that visited the Earth from outside. I believe it was erected by ordinary ancient Britons who were just as intelligent as you or I.

CARAS: What about extraterrestrial life? Do you have any information or does science have any information that militates against there being life on planets other than Earth?

HAWKINS: Throughout space there is intelligent life – intelligent organisms.

CARAS: What might their nature be? If their morphology is different from ours, which indeed I suppose it could be, but they were supremely intelligent (which we might assume them to be), would they be called men?

HAWKINS: I don't care what the organisms in space look like. I only worry about their thinking ability. As soon as the organism gets to the stage where it understands its environment and uses its environment then I would recognise it into the club.

CARAS: Would you care to speculate further what it might be like? Speculate, extrapolate on what its capabilities might be if, for instance, it comes from a planet very much older than ours? Instead of having a civilisation ten thousand years, it has one say four million and ten thousand years old?

HAWKINS: If you extrapolate the potential of life on the Earth for thousands of years ahead the potential is staggering. If you then wonder whether advances have been made in other planets in other parts of the universe you must conclude that there is every likelihood they are superior to our present status of development. Man on this Earth has only just arrived as a thinking, useful organism. For the past fifty years he has delighted in a technology that is quite new. In the next thousands of years he presumably will develop this.

CARAS: What might some of the capabilities be of a civilisation very much more advanced than our own?

HAWKINS: An advanced civilisation should be able to store factual material in its own body much more readily that we can. It should have almost total recall of information. Its logical processors should be superior to ours. Its understanding of the universe should be much superior to the understanding that we presently have on the Earth.

CARAS: You mentioned before the word 'computer', that Stonehenge was a computer. What about machine intelligence as it is here on Earth? Where is it going?

HAWKINS: The computer on the Earth is turned rapidly into a willing slave. At no time in the last few years since its introduction has it threatened to take over. It does not think yet like a human mind and I don't think it ever will. The solving of a problem requires tremendous imagination, tremendous searching through many stacks of blocks of information in the brain and at the present time it is totally impossible for a mechanical and electronic computer to do this.

CARAS: Do you think this will always be the case?

HAWKINS: Well, the future for computers is difficult to predict but I am sure that we will never build computers that get out of hand, computers that tell us what to do, computers that tell us answers that we didn't expect, because there is always a switch and man will always be making the computers and if he doesn't like what's going on I'm sure that man will throw that switch.

CARAS: As computers become more complex, as we assume they shall, is there any possibility of their developing personality disorders?

HAWKINS: A computer can develop personality disorders. In fact some of the troubles with computers today are very similar to the problems that disturbed people have. Of course computers are put in a sort of dream sequence even now to clear out their memory and recycle them back to some clean starting point. The analogy between computer disorders and the disorders in the mind of man is a good one. But I am not worried about computer disorders because when a computer goes wrong we can junk it. When a human being goes wrong we have to try and do something about it.

CARAS: What about awareness of self? Is this on the cards? Will there ever be awareness of consciousness?

HAWKINS: I don't believe that a computer will ever have the awareness of its environment in the way that man has. I think we need a living organism. There is something different between the living and the dead.

CARAS: Will that boundary line ever become cloudy?

HAWKINS: The boundary line between animate and inanimate material will always remain the same. If man is making equipment of a mechanical or electronic nature this equipment is always under his control and he will never feel the sadness at ending the 'life' of what he has built. There is quite a difference between organic life and inorganic objects.

SCENE TWO / TAKE TWO

CARAS: Doctor, what is the importance to science and to philosophy in our making contact with extraterrestrial life or at least becoming aware of its existence?

HAWKINS: As a scientist I would say that it is essential for us to establish some contact with extraterrestrial life. I personally would be very surprised if by the year 2000 we haven't received some messages from outer space. If we don't receive messages from outer space we are forced to the conclusion that we are alone or, even worse, that nobody likes us, and if we are indeed alone and intelligent life is unique, we will have one of the most difficult things that there is to explain.

CARAS: Are we doing enough now in the search for extraterrestrial life?

HAWKINS: We have carried out some searches but the search could be intensified. We could monitor the obvious stars more systematically. We could, for our own part, send out a message into space. It would take years and years to arrive at its destination but I would very much like to see a high-powered intelligent message be deliberately sent from Earth to certain stars that I could earmark.

CARAS: What should the message consist of?

HAWKINS: Well, the message that we could send could be sent very quickly by binary code. I would suggest, not to be facetious, but perhaps the first volume of the Encyclopaedia Britannica could be encoded and sent out or, getting down to more mundane facts, simple arithmetic. Arithmetic numbers: 1; 1, 2; 1, 2, 3; etc. Algebraic operations. A little time capsule could be made that could be sent out into space and it would be travelling over the immense distances for years and ultimately would be received. I think that for the future we should invest in sending a message right now.

CARAS: Do you think we are most likely to become aware of extraterrestrial life by a message deliberately sent toward us or by a simple monitoring of radioactivity on the surface of the planet or from that planet to another planet nearby?

HAWKINS: I don't quite understand your question.

CARAS: Well, do you think the first message, the first radio signal and code message we will pick up will be one that is deliberately aimed out towards unknown bodies – namely, us. Or would be just monitoring of radio activity?

HAWKINS: Accidentally, you mean?

CARAS: Yes, accidentally. Are we just going to hear a commercial?

HAWKINS: We will in the next decades receive some signals from space if there is any highly developed organism out there. The first messages that we receive must surely be directed at 'to whom it may concern' because to send signals accidentally and at random is a very inefficient process. I would imagine that anybody trying to communicate with us would single out the Sun and would produce a very narrow beam in their signal so that a tremendous amount of energy could be transmitted towards us. I would imagine that the first messages that we receive would be definitely meant for us.

CARAS: Might we not speculate and think that some civilisations within the universe, the cosmos, are very much older than we are and are therefore very much more sophisticated? Is it possible that we are being observed by means not yet at our disposal but at theirs?

HAWKINS: It is quite possible for another civilisation on a planet or a nearby star to be eavesdropping on our TV signals, radio signals. It would surprise me though if eavesdropping were to continue for long without a positive effort made to contact us directly, so I doubt whether eavesdropping has been carried on at present. You have to remember that we started sending out TV signals in 1945, say. Well, twenty years have elapsed. Those signals have only gone twenty light years into space. In our estimates of civilisations in space we estimate that a search will have to be made over the volume covered by a hundred light years, not just twenty. So even since the invention of TV high-powered TV signals have only gone out one fifth of the region of space that we are presently interested in.

CARAS: If an extraterrestrial intelligence should get our signals and hear about the Korean War, the Vietnamese War, the Russian-Sino conflict, might they be very anxious? Would they perhaps not be at all anxious or have anything to do with us until our signals started to deal with matters more in keeping with galactic war and peace? Would we not, from our radio signals if they were monitored, look terribly hostile and not a very good neighbour to take up with?

HAWKINS: A civilisation on a planet going around Epsilon Aridani is more than ten light years away from us. They would eavesdrop and pick up information about the Vietnam war in 1975 at the very earliest. If they had an antidote and could tell us how to correct the problem the signal would arrive here in 1985. So the possibility of help, direct intervention with the problems that we have right now, is totally out of the question.

CARAS: That is a good point. Also the point, would a civilisation monitoring us accidentally, even twenty years from now, would they want to have anything to do with us? I mean, after all we must appear to an advanced civilisation that have got this out of their system, we must seem like an awful savage group of creatures.

HAWKINS: I cannot speak for the civilisations that we believe to exist up there. What they are thinking is their own business. Their motives are their own motives and until we establish contact, and we sincerely hope we will in the next decades, until we establish contact we cannot be sure of their thought processes and their, if I may use the word, 'human' motives.

CARAS: Is there anything else further on this general area? Any further statements or thoughts on extraterrestrial life specifically?

HAWKINS: Yes, one. What is the universe designed for? Because if we don't find life then we have again lost the answer to the question, 'What was the universe designed for?' If we expect life to be scattered throughout the universe you can reasonably argue that life is important. If there is no life then the stars are important and the galaxies and what is man in that set-up? That's the feeling that I would like to express. That there is an urgency in establishing that there is life elsewhere in the universe because a dead universe certainly is without connection to us. I do not think that we could argue that the entire universe was ours. It's too great and therefore we don't rate in it.

SCENE THREE / TAKE THREE

CARAS: Doctor Hawkins, would you like to rephrase the statement you made earlier…?

HAWKINS: Yes, I would like to scrub what we said about Vietnam. I think it has nothing to do with the situation. The question is raised, could we get any help from outer space? I would put it this way: that the time delay is going to prevent any help coming to us right now. Even if there were a civilisation that could eavesdrop on us here, listen to our problems, translate our language, and learn to communicate with us instantly, it would take ten years for any of our TV signals to reach the nearest possibility for habitation. If they could magically communicate with us instantaneously it would take ten more years for the message to get back, so a very minimum would be twenty years' delay on any advice on any problem that we have right now. So I rule out of the question any help for our present predicament.

CARAS: The thought of intelligent extraterrestrial life seems always to take the promise far out to other planetary systems. What about life, not necessarily intelligent life, what about life within our own solar system other than on this planet? What may we find on Mars? We will be there fairly soon. What may we find there?

HAWKINS: There's nothing interesting in our own solar system. Venus is out of the question, it is too hot. Mars has a very bleak climate; it's as though you could take the plateau of Tibet and raise it even higher than it is now. If there is anything surviving on Mars it is going to be the most miserable form of life you can imagine – some little blob, some little mushroom type of thing which would have no interest to us. Life on Mars is certainly not going to communicate with us at any time.

CARAS: It would, however, communicate with us in another way. It would give us facts that life is not exclusive property. What about cultural shock? What might the reaction be other than in scientific circles, where there would be elation? But what would the general reaction be do you think to contact with an extraterrestrial intelligence?

HAWKINS: What are we going to do when we get messages from outer space? I have debated this in my mind quite a lot. On the one hand I can imagine a scientist saying, 'This is it. We are receiving messages from outer space,' and the general public might be so well educated that they might say, 'So what?' and they might regard it as perfectly obvious. On the other hand, and you can never predict public reaction, we may have the biggest panic that we have ever had, the realisation that our little planet cannot just be regarded as an isolated neck of the woods. This realisation may indeed shock the whole population of the world, and if the population is shocked, goodness knows what might happen. We might find that it is senseless to be fighting over national boundaries, we might make great strides and it might be beneficial.

CARAS: People predict a lot of UFO sightings. There has been a great deal of argument about manned space flight versus instrument probes. What are your feelings on this as an astronomer?

HAWKINS: As an astronomer, I am nervous about man going into space. I would much rather see mechanical instruments which if we lose them, there is no sadness involved. Each time an astronaut leaves the Earth I am involved personally in his flight, like most people, and I am dreading the time when we have to face a loss of life in space. I would much rather see us move without risking life until such time as we were really sure of what we were doing.

CARAS: I don't want to seem harsh and cruel but this argument you've given is off and on. On the other hand one says we shouldn't have window cleaners scaling the Empire State Building. We do build high buildings and men do ride up girders: we do go to races and see drivers drive a 110 miles an hour. We think nothing of our sons or brothers going into the armed forces and into the parachute corps and jumping out of airplanes and we are sending troops to Vietnam and all these endeavours, some of them far less significant surely than the exploration of space, and yet someone says that we are not to have manned space flight because we may lose two or three astronauts. Is this really a factor?

HAWKINS: I personally am nervous when astronauts go into space and I personally would prefer mechanical equipment to go. Yet I am excited when they make the trip. I appreciate the risks they are taking and there is no doubt at all in my mind that these people – each one individually, is a hero and so I am acting as a very conservative person expressing great appreciation for the risks that these people are taking.

CARAS: Do you think that instruments can ever be perfected that will do the work of an astronaut? Have judgements, know what to look at? Know when to stop and look longer than they would? I am curious about them.

HAWKINS: Please don't think I am saying man should be Earth-bound for ever. The challenge of the cosmos is in front of us and there is no turning back. We must explore. The rate of exploration is the question that worries me. I personally would never, even if it were made completely possible for me, I would never step into a space vehicle because of the natural nervousness that I have. But we have to face the challenge ultimately and the rate that we face it is the responsibility of the people in the space programme.

CARAS: Colonisation… Surely we will do more than just touch down on Mars? What are the potentials for colonisation? For self supporting cultures that have flown from Earth to exist elsewhere?

HAWKINS: I think there's no reason at all for us establishing colonies anywhere in space. The problem of overpopulation on the Earth should be handled here. The breeding capabilities of man combined with woman is so great that we could populate almost all the planets within a feasible range so I don't see any purpose in colonising in the sense of the word that colonies were set up in the past. The materials in space are not worth considering for mining purposes. If gold existed on the surface of the Moon in solid blocks the cost of going there to bring it back would far outweigh the value of the gold.

CARAS: What about scientific stations? Wouldn't you like to see an observatory on Mars permanently established where it would not be mucked up by an atmosphere?

HAWKINS: The only reason for taking Homo sapiens and allowing him to live elsewhere in space is the exploration aspect, the scientific aspect. A station on the Moon would be invaluable to us. A station on Mars would be equally valuable and the scientific results that would come out of this would be enormous.

CARAS: Do you suspect people will ever be born on these planets?

HAWKINS: It's a possibility that people could be born on a remote station. I think it would be an accidental circumstance. I don't think we will be having honeymoons on the Moon as Dr Whipple suggested.

FORTY YEARS ON: Gerald S. Hawkins, 2003

In 1965 the space programme, beginning with Sputnik in 1957, was inspiring the whole world, and we were only four years away from landing on the Moon. Stanley Kubrick's film was not released when my interview took place, but I was well prepared for (Sir) Arthur C. Clarke's imagination. Back in London in 1950, after an astronomy conference, he watched me putting on my motorcycle equipment ready for a long, cold winter journey to the Jodrell Bank Radio Telescope. He soliloquised: 'As Dr. Gerald Hawkins zipped up his space suit, little did he heed the tell-tale hiss of gas that was to spell his doom on the Red Sands of Mars.'

Not that Arthur's words influenced my outlook on travel in the Space Age. I was, and still am, dedicated to increasing our understanding of the universe. But I was against the idea of men and women in space (at this stage of our development, that is). As I said back in the 1966 interview, I was dreading the first loss of life. The science fiction of *2001: A Space Odyssey* made space travel a glorious adventure, but when the real year 2001 came, we knew that the death rate had been established as a shocking one in fifty.

After forty years of exploration, instruments have given us more information about the Earth and the cosmos than manned flights have ever done. As an astronomer, I know that beyond the thin onion skin of atmosphere wrapped around the Earth, the greater part of the universe is totally hostile to life. Like the gods told King Gilgamesh in 3000 BC: 'Remember, you are a mortal, enjoy your limited life on Earth.'

I'm glad that I supported the intellect of ancient peoples in the interview. They were not those hairy ape-men depicted in the film who threw the twirling bone into the sky! All the evidence I have gathered shows that the Stonehenge people of the late Stone Age had an expert knowledge of the movements of the Sun and Moon, and built their mammoth structure in southern England with an amazing technical skill, and without any help from ETs. It was not until the film *2001* was almost complete in the late 1960s that the Sun–Moon connection was discovered. The long delay of recognition was partly due to the pattern being skewed symmetrically about the midsummer sunrise, instead of the north–south line.

Arthur, as I knew him in the 1960s, shared my interest in Stonehenge, and I wonder if the idea of the monolith on the Moon in his story of AD 2001 came from those standing stones of 2001 BC?

At least one thing I said in 1966 has come true: any life on Mars is going to be in the most miserable form, with no intelligence we can share. Even if the particles in the meteorites plucked off the snow in Antarctica did originate on Mars, they are like the organic 'blobs' I predicted, hardly qualifying for the name life-form.

The interviewer, Roger Caras, did not tell me about the now famous cry of the stranded astronaut: 'Open the Pod Bay doors, please, HAL. Open the Pod Bay doors, please, HAL!' I was being asked surreptitiously about artificial intelligence, and whether a computer could develop free will. I foresaw viruses and freak, hacker-induced behaviour, which became the great scourge of the Internet. However, electronics, though speedy and complicated, I saw as a purely passive device. I still argue, as I did then, that a computer could not take over and act with a mind of its own. Even in the film, how could we say that HAL actually decided to shut the astronaut out? Computer HAL might have done what computers often do – crashed, or blown a fuse, and that to my mind was the technical reason for the Pod Bay door staying closed!

My deadline of *2001* for receiving communications from extraterrestrial civilisations has passed without any radio or laser-beamed signal coming through. In my book, *Mindsteps to the Cosmos* (1983) I hopefully extended it to AD 2021, but the situation is growing to be a serious one for the understanding of our place in the universe. A universe devoid of life raises the big question about the very existence of *Homo sapiens*, and has deep scientific and philosophical consequences that have yet to be faced.

Hope and expectations are the fuel of modern culture, and *2001: A Space Odyssey* projected those hopes right to the end of the twentieth century. But it addressed only the technical aspects of the new frontier of space. During that period things were not so encouraging back home on Earth, and when the real AD 2001 came along, human nature in sociology and geopolitics was exhibiting some disquieting aspects. Back in 2001 BC the Stonehenge people left a legacy of their knowledge. They lived much closer to the environment and seemed to share a purpose in their henges and stone monuments. What they were doing cannot be understood today by a simple walk-through. Their achievements were made the hard way, despite the harshness of their existence.

By AD 2001 many parts of the world had living standards undreamed of in ancient times, but it remains to be seen how humanity lives with its problems if these continue and increase, and what legacy will be left of our achievements in the years ahead.

Moonwatcher, the ape-man, tossed a twirling bone in the air, and it became a wonderful spaceship. We have launched thousands of satellites into orbit around the Earth and into the solar system. A disc on the space vehicle *Voyager* carries samples of our music and a binary code message far beyond the orbit of Pluto, onward to the nearest stars. These artefacts will survive in space to show our mechanical achievements almost indefinitely. No doubt in the future these twirling things will be regarded as one small technical step, and hopefully, when 3001 finally comes to pass, humanity will indeed have made its 'Giant leap for Mankind.'

Francis J. Heyden SJ

[Professor of Astronomy, Georgetown University]

6 MAY 1966 / SCENE ONE / TAKE ONE

CARAS: Father, do you believe that there is a chance of our discovering extraterrestrial intelligence in the foreseeable future?

HEYDEN: I do not know whether I could say if it is the foreseeable future but I wouldn't be surprised if we did not, some time after we develop lasers and radio astronomy a little further. Maybe our techniques of recovering information by statistical methods, which out of all these observations we make in space, pick up some signal that is from an intelligent source.

CARAS: From a scientific point of view strictly for the moment, is there anything that militates against the existence of extraterrestrial life, even intelligent life?

HEYDEN: I think that there is nothing that militates against the discovery of extraterrestrial life. My reason for saying anything like that is this: first of all, we are just one planet in a little corner of a great Milky Way that is only one of probably a billion galaxies. Why would all the other material be there as far as we are concerned? Here we'd say, in my profession, that it gives glory to God but, as a scientist, you would say there must be some others like ourselves somewhere.

CARAS: How would you suggest that we go about looking for them?

HEYDEN: Well, we would go about looking for them by means of the scientific devices we have at hand at present, or anything new that would be found in the future. For example, quite a while ago, over a century now, some explorers found a stone called the Rosetta Stone and with the little inscription that was on it in two languages we were able to break the cuneiform of the Egyptians and be able to read that language from just a few words. Now, in the same way, we are picking up all kinds of static at present on radio telescopes which we call noise. Much of it is man-made noise, probably from this planet. People with electric razors, automobiles with unmasked spark plugs and so forth, but out in space too there is a certain amount that we understand, that we know that it is produced by various molecules, or interstellar hydrogen. But among that also is other information that some time in the future we might find it to be, just here and there, a little bit of an intelligent signal that is being broadcast. However, when we learn more of the use of lasers for communication it is quite possible that in the spectra of some stars some time in this we might find an intelligible signal. I am going way out now when I say something like that because we have not arrived at anything of that sort yet.

CARAS: Does the existence or probable existence or theoretical existence of extraterrestrial life, particularly extraterrestrial intelligence, create philosophical or theological problems? Is there anything in the scriptures and is there anything in philosophy or religion as you understand it that militates against this?

HEYDEN: I see nothing in either religion or philosophy that would militate against it. In fact, as a philosopher, I would say this is something we must investigate because it is the role of the philosopher to study all things. Now, as far as religion is concerned, I do not know of any eleventh commandment that says: 'Thou shalt not talk to anyone off the face of the Earth.' It is a matter of fact though this would be a big problem. I can understand where we would receive a message from someplace out in space but to talk back to them and exchange ideas, that I would say is close to the physically impossible. Physically impossible for the present and I will tell you why. It is because all our communication is done by means of electromagnetic radiation. So that even if someone on a planet near the closest star were to send a message to us it would take four and a half years from the time he said 'Hello' there until we heard it here and by the time we answered, 'Glad to hear from you' and 'How are you?' it would be four and a half years before he got our answer. So you see, it would take quite a lifetime to just have a very casual passing of the time of day.

CARAS: If, in its morphology, if we encounter extraterrestrial life, it looks nothing like us at all, it is a totally different form of animal, from the point of view of theology, this is still man? When is it man? When is it not man in spite of its morphology?

HEYDEN: Well, if it doesn't look like man at all and it is intelligent and we could communicate with it and exchange ideas, I do not know whether we would call it man or not. We don't call angels men but all through the Old Testament we have had constant references to certain angels who are pure spirits, who have been messengers of God to men. Other intelligent beings I would say, we would call this at least an intelligent being like ourselves. The only thing I hope if we ever meet something that doesn't look like man coming from outer space, even if it is intelligent, I just hope it isn't hungry!

CARAS: Then there are no apparent contradictions between what we believe and how we should react and the fact of extraterrestrial life?

HEYDEN: I have never heard of any and I don't think there are any. There is nothing in revelation and certainly nothing in the teaching of any church I know of that would be against it.

CARAS: Is there anything you want to say further on this? Something before we go on?

HEYDEN: On this question of the meaning of religion in exploration of life outside of our own Earth, whether that life be on one of the planets in space, or whether it be on even some star in a solar system all of its own, we would only have one question that would come up and that would be the question of the fall of Adam and Eve and the inheritance that we have gotten from that fall. These people probably never had an Adam and Eve that ran into trouble in Paradise and got thrown out and possibly they are in what the philosophers and theologians of the Middle Ages referred to as the state of pure nature, meaning that they do not have the darkened intellects and the weakened wills that we have and therefore if we met them we would probably meet some of the finest guides and consultants and intelligent people that we could ever run into. It would be most beneficial for us. On the other hand, if the fall of Adam and Eve had been repeated, we might meet some people that would benefit by the same fruits of the redemption that we do, would understand us, speak almost the same language, but this is all speculation. We do not know anything about that. This is the only thing I see in all the religion that I have studied through courses on theology that would raise any question of this sort. And it still would be something I would have to solve in this particular instance and which has not been defined by any matter of faith and morals.

SCENE TWO / TAKE ONE

CARAS: Father Heyden, a being from another planet is no less a creation of God than maybe one on this Earth?

HEYDEN: That is definitely true, that a being anywhere in space is a creature of God. So that if you even took some living thing that would be on the most distant galaxy it would be a creature created by the same god that made us all. Now, I have a scientific reason for asking anybody to believe a statement of that sort, because when we look at objects in space, with the instruments that astronomers use, especially a spectrograph, we see in the spectrum the lines that are made by the electrons at different energy levels and the various chemical elements that we are familiar with here on Earth and in order to have science advance the way it should, that is seeking after the truth in a material world, we have to postulate that we have the same physical and the same chemical laws throughout the whole universe. In some respects I suspect that eventually we may know too that we have the same biological or vital laws throughout the whole universe. So hydrogen in the most distant galaxy is behaving exactly the way hydrogen would be behaving in a laboratory in any observatory or physics department of a university on this Earth.

CARAS: God has manifested Himself to man in many ways, including in the form of His own Son. Would He have manifested Himself to other forms of life, if such existed on other planets, and how would He have manifested Himself? How would Christ have appeared to an inhabitant of another planet with a totally different morphology?

HEYDEN: God can be known by any intelligent being. It is the belief of the philosophers from Aristotle and those who preceded him even that an intelligent man looking at the wonders of nature, studying them, either from the standpoint of science or admiring them simply from the standpoint of beauty, with order and variety, to them will come the realisation that there is a supreme being, or a God. On the other hand we know from the Bible that God has revealed Himself to mankind on this Earth in a very, very special way in order to help us appreciate Him more and more, or to come to know Him better. Now, as far as other people in space are concerned, if they are intelligent they will know God the way a scientist would know him, or the way Aristotle and other great philosophers down through the ages have instructed us that we can argue from the conditions of things in this world, to the existence of a supreme being. That, of course, is that since we are intelligent and have personalities about us we would understand that God, who made us, would have to be intelligent and have personality, though we would not know whether these people have as full an understanding of God as we have through revelation until we found out what form of revelation was given to them.

CARAS: You said before – I think you did it as an answer to my question and I would like to have a clear statement – that you know of no contradiction then in the philosophies and theologies with the concept of extraterrestrial life?

HEYDEN: There is no evidence in all of revelation that there can be no extraterrestrial life already further than that which you find on this Earth. There would be no reason for putting that in revelation, really, because if there were not any kind of life and revelation wanted to tell us that, I do not know what good it particularly would do us, except perhaps it might save us from some investigations and studies which we might look for. But, in looking for extraterrestrial life – I was going to say interstellar lives – we are not going to do this probably by a direct search simply for that. I believe that it is going to be something that we will find, in the course of other investigations. There would be a by-product.

CARAS: What could man learn from the experience of contacting extraterrestrial intelligence, if it exists?

HEYDEN: What could man learn from contacting extraterrestrial intelligence? You know, that is a very interesting question for us to think about. First of all, he might know more than we do. We might get answers to a lot of things, but, whether we could contact him directly and not wait for the long time-lag between question and answer that might be as long as a couple of hundred years. On the other hand, he might be just giving out information and as we gathered it in and interpreted it – why, we would be advancing our own reservoir of knowledge in any field that he happened to be communicating to us. In addition to that, though, I would say this: that the satisfaction of knowing that there is someone else out in space, who is able to communicate by intelligent signals, is a great advance in our own knowledge – and helps us, I think, to appreciate the position that we have in this and our own role in the universe.

CARAS: Father, before, when the camera blew up, you were making a statement about there being no conflict and, you were making it without any reference to denying it when I asked you. Would you mind doing that again so that we have it clear?

HEYDEN: It is sometimes suggested to me that possibly there is a conflict over investigating whether or not there are intelligent beings somewhere else than on this Earth. People very often feel that we have been put here on this Earth to stay and that is where we are going to stay, and that space is God's locker – we are not supposed to pry into it. I do not believe that. After all, we do have the psalm that says 'The Heavens tell of Thy Glory, O God' and also, as a philosopher, we have the urge to learn as much as we can. We have restless minds as well as bodies. Minds that are always enquiring into things, to learn how they are made, how they work, what they are and there is nothing in all of revelation that is against learning more and more. As a matter of fact, I think myself that the man with what I call 'scientific' faith is much more at rest in his soul than one who is just satisfied with what he knows from reading scriptures.

CARAS: Is there any significance to the fact that apparently for a very long time cultures of wide diversity have always referred to the abode of God, His being in the Heavens and the Heavens is not down, it is not in a cave, it is not in a rock, it is up. This seems to be intuitive, instinctive. Is there anything in this of significance to us, in contemplating space and is there the possibility of dwelling or dwellers in space?

HEYDEN: There has always been the attitude among believers as far back as we go in the history of religion that God is in His Heaven, God is above. One reason for that, of course, is that we are conscious that we don't see God face to face here on this Earth and we have been accustomed to the beauties of, let us say, a sunset. If you stand looking at the sun as it sets, this beautiful panorama in the sky, something like Shelley's 'In the golden glory of the setting sun, or with clouds a brightening' – and so on – I can see where a person might want to make that a sort of a shrine for himself. I can see and understand the idea of a sun-worshipper just from that and I think that it is because of that that we are inclined to place the residence of all divinity in space or up in the Heavens above because there is so much beauty and so much wonderment there that we do not understand. But again, as one having scientific faith, I know that God is everywhere and that God is as much in my little finger as he is in the most distant star.

SCENE FOUR / TAKE ONE

CARAS: There are a lot of hysterics, Father Heyden, and there is a lot of talk about recent visitations from outer space. Do you see any significance, philosophical significance, in the fact that the Venusians and the Martians or the Saturnians or the Plutonians are generally described as very benevolent-looking souls in long white flowing gowns with white beards and aquiline noses?

HEYDEN: There is the question of what would a person from another planet, let us say in this solar system, look like? Perhaps it all depends on who is trying to describe such denizens of other planets. If it is a cartoonist, generally the Martian is a little bit of a rat with an egg-head, a couple of antennae sticking out of the top, but then there are these other people who claim that they have not only seen flying saucers land, but have been aboard them, have talked with the people who were on them and, in every instance, they talk about how these people are very much like ourselves, that they were very benevolent, very kind, very intelligent, seemed to be very, very advanced. Well, of course, I don't believe myself that anyone has ever met somebody who has gotten off a flying saucer. And this then would be what I would call wishful thinking of a sort. As regards the way they might look, I have an idea that people who are intelligent beings on another planet might be very much like ourselves simply because of the uniformity we find in all the physical and chemical laws of the universe. Living cells that came on to this planet somehow might be the same kind of living cells that came onto another planet somewhere out in space and that the same laws of validity that we are trying to learn now might be the same ones that governed them. And that is quite possible. That you wouldn't just tell the difference between a living cell here and a living being from the Earth or one from outer space. That, of course, is pure speculation. I do not know the answer to that.

CARAS: The carpet-hysteria, the spate, and one hears very often from a very conservative person, who says that if God had intended us to travel in space He would have put us there, or that had God intended us to fly He would have given us wings. This fool type of emotional reaction to progress, what is your reaction?

HEYDEN: There is a great conservatism among some people that we are not supposed to pry into, let us say, the mysteries of space. This is, of course, against all of the efforts of astronomers from time immemorial who have tried to know more and more about the stars and about the planets and the laws of their motion. What a tremendous achievement it was when first of all Kepler found from observational data the three laws of planetary motion and having approached his goal from observational data he was followed about two generations later by Isaac Newton who started out with a theory, and from his theory, derived the same three laws that Kepler had gotten from observation. This is real scientific advance and I don't think anyone in the world is ashamed of it. As a matter of fact, putting a satellite in orbit about the Earth in this generation, and I think we do live in a wonderful generation of discovery, was an achievement of a dream that Isaac Newton had. Isaac Newton suggested that we put a cannon on top of a high mountain and fired it parallel to the direction of the gravity of the Earth at such a speed that the cannon ball would go in orbit about the Earth. No one would say that we have done wrong by means of artificial satellites because we have actually achieved something that is worth all the money that we have put into it at present. The weather satellite, communications satellites for example. Now if we are to go further in our investigations of space I do not think that we are going to violate any law of God. I don't think any conservatism on the part of people should have any influence on us and I will tell you why I think so. There are many people in this world today who are quite conservative in their attitudes towards space exploration. Some think that they don't have any need for all of this, that there will be absolutely no benefit that will come of it, that probably we are just going to waste money and human lives in the long run. Nevertheless when we look over the record we have probably achieved something new on our weather satellites and communication satellites that is worth all the money that we have put into them, so far, for this investigation. And in the beginning we were not particularly thinking of anything like this. What we were really thinking about was learning more about the mysteries beyond the atmosphere of the Earth. Learning something about this cosmology. Of how this solar system of ours was put together. Did it all come into being at one time? Or is it something that has developed over the billions of years from the aggregation of little grains of sand and dust and gas from out in space among the stars until we have this beautiful mechanism that is moving as it is now like a watch?

CARAS: The theory that life did arrive here in spore form, is this in contradiction to religious precepts and teachings?

HEYDEN: The origin of life on the Earth is a question that I do not know if it is ever going to be solved. On the humorous side we have some people who think that it evolved from some garbage that was left behind after the crew of a spaceship stopped here one time, had a picnic and then departed for another planet. Other people believe that life started after a molecule had been developed to a certain degree of complexity, so that it would support life such as DNA and the conditions for life in the temperature and other chemical requirements such as oxygen were found and then, somehow or other, a pre-life germ or some sort of thing started it going and you had your very simple cells that gradually became complicated animals and eventually, out of it all, came a body of an animal that God decided to give a human soul and call man. This is all right if you want to believe in that form of evolution as long as you remember that the human soul that we have is not something that evolved out of a chemical test tube, or whatever you had in the early stages of development of the Earth. We know that life did not always exist on this Earth. We know that life – living forms have changed and do change over long periods of time on the surface of the Earth, but all we do have is this one brief history of Homo sapiens, or the intelligent man, and we do not know exactly where he came from, how long he has really been here, but we do know that over the period of written history, we have seen a vast development in his knowledge but not so much in his intellectual abilities. I think the world is going to wait a good many centuries before it finds another Aristotle or an Archimedes and I know that there are people in this world today who are very intelligent people who really and sincerely believe that a man of that calibre and intellectual ability would not need very much time on Earth today to catch up with all of the latest theories and scientific knowledge we have.

Norman Lamm

[Associate Rabbi of the Jewish Center, New York; Associate Professor, Jewish Philosophy, Yeshiva University, New York]

10 MAY 1966 / SCENE ONE / TAKE ONE

CARAS: Rabbi, is there anything in the philosophy of the Jews that militates against there being extraterrestrial life?

LAMM: In Jewish thinking throughout the ages there has been a general tendency to regard man as the sole concern of God in this world. Accompanying this there has developed an idea that man is the purpose of the entire universe. This has been a rather general tendency, although it never has been enshrined into dogma. However, one of the very greatest of all Jewish thinkers of all days, perhaps the greatest Jewish philosopher of all times, Maimonides – about eight hundred years ago – strongly opposed this view and he maintained that man may be the superior creature on Earth but it would be foolish to consider that he is, therefore, the purpose of the universe or, in fact, that he is necessarily the most advanced creature in the world. And according to his whole approach, Judaism today can welcome with remarkable openness the idea that intelligent races exist elsewhere, species even more intelligent than man.

CARAS: If there are intelligent races and they have had millions of years longer than we have to develop, what would happen if they are not men? What would happen to the theological position, or must they be known as man?

LAMM: I think the problem as to what form these extraterrestrial intelligent sentient races take is really irrelevant. They may look like elephants, they may look like seals, or look like the little men from Mars that science fiction has always portrayed in the comics. The point is that it is not morphology or form that counts. What does count is their intellectual and spiritual development. If they are more intelligent than man, then I suppose they will be less hostile than man and that therefore we can enter into some dialogue with them - assuming, of course, that we ever can contact them.

CARAS: Do you think we will contact them?

LAMM: I am not a scientist and I think even the scientists don't have any definite answer. They can have an educated guess and measure probabilities, but no one will know for certain until we hear that first voice.

CARAS: Do you, as an individual, believe that there is extraterrestrial life?

LAMM: As an individual I think there probably is intelligent extraterrestrial life, but that is not something that would displace my central concerns. My concerns are and should primarily the perennial issues of human existence. I think there is a great deal of truth in the humorous question, 'Do you think there is intelligent life on Earth?' We have our very serious problems – here on this planet – the kind of problems that may cause us to disintegrate before we ever get a chance to find out if there is life elsewhere. So, while I certainly would not discourage or in any way denigrate the search for life elsewhere I think that it's a fascinating intellectual problem and that human curiosity certainly must be satisfied, but that the overriding problem is how to live in this world and live decently and well.

CARAS: Scientists have pointed out a number of times that if there is intelligent life in the universe and I think most people today tend to believe there is, outside the Earth, that it would be intelligent life in many cases millions of years older than ourselves because, simply, we're a fairly young planet around a not terribly old star. They speak of what we could learn from such creatures. Technologically our own civilisation is ten thousand years old and yet they may have a civilisation that's four million, ten thousand years old. What might we learn from such a race? Speculating from the point of view of theology and philosophy, surely there are mysteries in our theology? Mysteries in our philosophies? What if we had four million years of experience more with it? What might they teach us about God?

LAMM: I believe that if we should ever come into contact with an extraterrestrial highly intelligent race much older than our civilisation that the primary lessons we would learn from them would not be technological. If it's merely advanced technology then it's nothing but another human race, even though it may be shaped differently, but not qualitatively different. If there is a civilisation which has lasted that long, then by the very fact of its survival it has indicated that it has learned certain secrets of social justice and harmony and the ability for individuals to get along without killing themselves off on a large scale. I believe that will be the main thing we can learn from them. I don't think that theological insights are cumulative, that mere age or number of years of experience gives one more or greater theological insight. It may give him more technological knowledge, because technology is cumulative, but theological insights come more in jumps and leaps and depend upon individuals without necessarily being additive. Perhaps they may have more knowledge than we have theologically. I am committed very deeply and very personally as a believing Jew to the idea of revelation. This does not preclude a revelation to intelligent beings elsewhere because the Jewish belief is the Biblical teaching that man was created in the image of God. This means that man shares something with the Creator. This sharing may be interpreted as intelligence, as ethical inclination, as creative ability, and there is nothing in our tradition that insists that man is exclusively the being who shares these faculties with his Creator.

CARAS: Some scholars somewhere pointed out the strange correlation that since time immemorial the concept of Heaven has always placed this heavenly abode up, not down. Not in the core of a tree, not inside of a rock, but up. He has attempted to correlate this with the belief that Earth may have been visited in the past by extraterrestrials, a very long time ago who drew to themselves and to their position of ascent physical...

SCENE TWO / TAKE ONE

CARAS: We were talking about Heaven being up.

LAMM: Yes.

CARAS: Is there any correlation that can be implied between this and the thought of extraterrestrial life?

LAMM: The fact that most religions have located God as residing, as it were, in Heaven – I don't think has any real relevance to the possibility that at one time extraterrestrial creatures visited the Earth by parachuting out of Heaven. I know that in the Bible the term shamayim or Heaven as used for the dwelling place of God is meant purely metaphorically. The Bible begins with the verse that in the beginning God created the heavens and the earth. If God 'lives' in the heavens, where was He before He created them? For instance, the Bible says that the Lord spoke to you out of the Heavens and yet, at another place, it says that the Lord came down on Mount Sinai. Obviously, He couldn't speak from the Heavens if He came down. Of course, all this is meant as metaphor. When the Bible speaks of God residing in Heaven and when men generally speak of God being in Heaven, it is meant as metaphor, because when one looks upward one is overwhelmed by the vastness of space. Even Immanuel Kant maintains that when he beholds the heavens he is overwhelmed by the natural order of the universe, even as when he beholds man, he is overwhelmed by the moral order of the universe. There is no direct correlation between, as it were, the upness of the sky and extraterrestrial life. I will say, however, this: that in the Bible itself in the sixth chapter of Genesis there is a rather mysterious reference to the 'sons of God' who marry the 'daughters of Men,' and then a reference to something called Nephilim which means, as it has been translated, 'The Giants,' and whereas a good part of Jewish exegesis defines that as simply a race of strong men. There are certain Jewish sources, particularly Philo in ancient Egypt and Josephus, the Jewish general and historian in the Roman era, who saw in this a reference to a mysterious kind of race, which, I suppose, in contemporary terms would be described as an extraterrestrial race.[1] So that there are some kind of mysterious references to that sort of phenomenon, but of course it never has become any kind of official point of view.

CARAS: Mankind seems on the verge of developing a super-intelligent machine. This great-grandson of the computer will be with us in a few years. It's going to perform many of the functions, the mental functions, that men now perform. What problems lie in theology and what do you personally feel the future of this sort of thing is? When does a machine cease to be a machine? What if it feels pain? What if it is conscious?

LAMM: The problem of whether or not machines will become so great and so effective as computers that they will begin to replace man has no direct relation to theology, although it does have an indirect bearing. By that I mean that a computer, no matter how efficient, is fundamentally an extension of the human brain and this no more makes the human brain useless than the fact that the scissors are an extension of his fingers makes his fingers unimportant. It simply is a way that the human brain has managed to create machines to improve its own activity. It is inconceivable to my mind that a machine can be created which will be itself sentient, self-aware, and self-conscious. Of course, those who believe it have a problem. I do not think it is within the realm of real discussion, certainly not for the foreseeable future. However, should such sentient created machines also develop a moral sense, I would incline to treat them as a form of humanity.

CARAS: What about the machines themselves though? Not speaking now of

---

1 Genesis 6:4; 'There were giants in the earth in those days; and also after that, when the sons of God came in unto the daughters of men, and they bore children to them, the same became mighty men who were of old, men of renown.' (King James Version.)
The word giants here is the translation of the Hebrew Nephilim (from the Hebrew *napal*, meaning to fall, hence the fallen ones). The sons of God have been interpreted as angels, possibly fallen, but how could angels marry when they were reputedly sexless? Much has been made of this passage by the followers of von Däniken.

man being obsolete. I know you passed this over and we can skip it if you want, but they're now talking, well, Freeman Dyson for instance is, that machines will not be electronic, they will be biological machines that will have this superior capacity.

LAMM: I'll pass on it, even though, as I implied, the development of a moral sense is critical.

CARAS: OK, it's a problem.

LAMM: A naughty one, too.

CARAS: In the future many scientists, perhaps even most scientists, believe they will synthesise life in the laboratory. What happens here?

LAMM: The question of whether life will be synthesised is a real one, but not an overwhelming one. I believe that from everything I read that life will be synthesised in the near future in the laboratory. The problem, of course, is that Genesis teaches that God is the author of life. He is the Creator of man and of all living things. However, if we understand our religious beliefs correctly we will, I believe, discover that there is no real problem. Man was created in the image of and he was commanded to imitate God. The imitation of God is understood as being primarily ethical, moral. Man must be ethical because God, too, is ethical. In the same manner we ought to imitate God in being creative. Judaism maintains that man is the co-Creator of the universe, that God created the world unfinished, as it were, and gave it over to man with his intelligence, his instincts, and his technological inclinations, to complete the world and to improve it. To make it liveable and habitable. Therefore, if man can, as it were, imitate God technologically with lifeless stuff, there is no reason why he cannot be given the same grant of sovereignty to create life itself. This too is the imitation of God, provided, of course, that man remembers that it is his task to imitate but not to impersonate God. The problem arises when man becomes an impostor and plays God rather than imitating Him. Then he imagines that the world is his and he can destroy it even as he can build it. So, when man will eventually learn to create life, that will be a marvellous development provided man uses this knowledge beneficially. But the problem arises when man succumbs to his tendency to misuse his powers. When man begins to have power over life and death by creating life then the problems we face now in morals and in ethics, especially in the field of medicine, will become increased a hundredfold. The problems will then arise as to who is to say that the doctor, or the scientist, or anyone who has a PhD and is in control of certain processes, has a right to determine the future of anyone's, or anything's, life?

CARAS: Man is monkeying around with his environment. He is monkeying around with himself as well. He is changing himself. The Bible teaches, I believe, that God created man in His own image. Is there a theological problem here by which we will alter our own genetics? We will alter our own biology? We will purposefully create mutations in the years ahead?

LAMM: I cannot see any real problem in so far as the conflicts between man's creation in the image of God and genetic engineering, provided that genetic engineering is done under proper moral and ethical control. When we say that man was created in the image of God, we do not by any means intend any physical similarity between man and God. We do not maintain that God is some kind of invisible superman. We do assert that the image of God refers to His creative, intellectual, and ethical powers and, even more so, His freedom: freedom of will, freedom to do right or wrong, to do evil or good. Therefore, if man will exercise this freedom in determining the future course of his development in genetic engineering with moral orientation then that will be proper. In using his talents for creative and constructive ends, he will be fulfilling the image of God in the metaphysical sense and in the ethical sense.

SCENE THREE / TAKE ONE

LAMM: If we learn that from our contact with some kind of extraterrestrial intelligent race, far from us in the cosmos, that they have a concept of God different from ours, it is not going to count for much theologically. The reason for it is this: by the very definition of our understanding of God – as being far above nature, far beyond the universe and its Creator – he is essentially, in essence, unknowable. Well, if He is unknowable, that means that any other creative intelligence cannot fully fathom His essence. It matters little therefore whether this created intelligence is man on Earth, or other races elsewhere in the Cosmos. Their understanding of God may or may not be as advanced or more advanced than ours, just as there people and communities on this planet of ours who have differing conceptions of God. Some may be more primitive, some more sophisticated, and we still do not know who is 'right' and who is 'wrong.' Therefore any communication we have with other intelligences elsewhere may be illuminating, may be enlightening. It certainly will be something to consider. But I do not think that any created being has a monopoly on the knowledge of God. The knowledge of God is not something which is gained through technological advancements or even through intellect alone, but through spiritual insight. Unless they are well beyond us spiritually, which takes us to a different realm of discussion, I do not think we have any real grounds for either fear or undue optimism.

CARAS: The Bible, the stories and the legends, are full of references to men speaking to God, not just in their prayers, but speaking to God and God appearing to them. Not many people today claim to speak with God. They speak to God in their prayers but not with Him. Is it possible that in some other place, that some other intelligence's God is in more direct contact as, indeed, He is said to have been in the Bible with us?

LAMM: It is quite possible, that whereas on Earth the period of prophecy has ended and direct revelations of God to man have not occurred for many, many centuries, nevertheless on some other planet God not only has revealed Himself more directly but is currently in a state of direct dialogue with sentient and intelligent creatures. There is no reason, according to the Jewish perspective, why this should not be so.

CARAS: You talked before of – no matter what the morphology was – of an extraterrestrial intelligence. If it was a created being it was man, in the definition of man as he appears in Jewish theology.

LAMM: The importance of man will, or course, have to be re-investigated and I think reformulated, if we indeed discover that there are intelligent races elsewhere. When the religious traditions of our times speak of the uniqueness of man, it does not mean necessarily that man is the only kind of intelligent race in the universe. Speaking from the point of view of Judaism, if we should discover races of sentient and intelligent intellectual beings elsewhere, this will not destroy the concept of man's uniqueness but will expand it and we shall begin to see that man on Earth is a member of a very large cosmic community of bio-spiritual beings. So that these beings, no matter what they look like, whether they look like man or like little green animals or like snakes or elephants, nonetheless, if they are intelligent, if they are possessed of a spiritual dimension then they and we and all others throughout the universe who share this likeness to the Creator of the universe, and are part of a unique community, the cosmic fraternity of God-imitating creatures. If you believe in God as our great monotheistic religious traditions teach us, then He is a very big God - big enough to be concerned with all kinds of intelligent race, no matter where they exist throughout the universe. The fact that Earth may be only one small planet evolving around one small star, in a fairly insignificant galaxy does not reduce man's importance. Even if we are not in the centre of the world, I do not think that God is a social snob who will not be seen in the cosmic slums and alleyways. He is a 'great' God and a good one and even we on this little Earth are entitled to and receive His attention.

CARAS: The search for extraterrestrial life could consume generations, even hundreds of generations. Receiving a radio signal that may have been sent even a million years ago and sending it back out and receiving an answer, this is a long-term thing. This is not a short-term thing. How important is it to pursue this? If one of these radio observation programmes should receive a signal that is obviously artificially produced from an extraterrestrial source, how important is it to pursue it?

LAMM: It will be very important to pursue any kind of possible contacts with extraterrestrial life. Our natural curiosity as intelligent beings ought never be stifled and this ought to be expressed by a continuing search for extraterrestrial life, no matter how long it is going to take and no matter how many generations will have to wait for an answer. However, there is always a problem of economy and I don't only mean economy of money.

CARAS: How important is it to continue the search or to maintain the search or to pursue the matter once there are clues?

LAMM: The pursuit of any kind of information indicating some extraterrestrial race attempting to contact us is an important one, even if only to satisfy human curiosity, which is such an important factor in civilisation and progress, and even to satisfy us as children of God who want to know more about His vast creation and His manifold works. However, as I said, it is ultimately a question of economy, not only of money but of time and attention and, above all, talent and competence. It is not a matter of black or white, of either we do pursue it or not at all. The question is – at the expense of what else? My own feeling is that our problems on this Earth are so immense, so terribly important, that it would be wrong-headed to ignore them because of preoccupation with the efforts to contact outer life. This project will be so consuming of money and time and talent, that we shall be forced to neglect the minor league questions of mundane terrestrial life. We may then find that we are so involved in an attempt to contact some extraterrestrial race that by the time we are going to get an answer, there will be no human beings left on this planet to contact. Thus our first principle should be survival on our own planet. As an individual, I certainly am immensely interested to find out if there is life elsewhere and, certainly if there is, what kind of life it is, but not at any price.

CARAS: There has been an awful lot of hysteria over flying saucers, of course. There have been some hysterics who have been known as contactees – the ship came down from Venus. Invariably, almost conceivably, these Venusians, these Martians, these Saturnians, Plutonians, whatever they are, have long white beards, have white robes, aquiline noses. They are Moses-like, they are

Christ-like and they are Abraham-like. What does this tell us?

LAMM: I think that the so-called visions of the contactees are illusory phenomena. I don't put any stock in them. I am not as fascinated by the fact that they invariably, as you tell me, appear to be Moses-like figures with long white beards, white robes and aquiline noses, because on Yom Kippur, the holiest day of the Jewish year, both my grandfathers of blessed memory, when they wore their kittels, or white robe, looked just like that. I was very fond of them and I doubt if they were Venusian. I think they were more Galician – you know, part of Poland and part of the Ukraine.

CARAS: Anything you have missed you want to say on these subjects? Anything you want to pick up on? I think that we have covered our questions fairly well.

LAMM: Judaism is the kind of religion which affirms this world, the entire universe. We are other-worldly only to the extent that we believe that there is a kind of non-physical existence in the presence of God, but we do not therefore deny this world. That is why the Jewish tradition has always had such an affirmative and positive attitude towards science and technology when they are used creatively and constructively. It is in keeping with this tradition that I, as a Jew, am immensely concerned about the possibilities of life elsewhere. If this is discovered, this will in all likelihood make man feel more humble, but it will not humiliate him. On the contrary, those of us who have a religious commitment will feel that our horizons, religiously, have been expanded by discovering that God is greater than even our most profound theologians and thinkers have ever imagined Him to be. 'For the Lord is great above all His works' is a verse from our Bible that holds true now and that I suspect we will learn is even more true as our knowledge of the cosmos advances and progresses.

Checking the art department's model of the 'centrifuge' set

Sir Bernard Lovell

[Nuffield Radio Astronomy Laboratories, Jodrell Bank]

LOVELL: The Earth on which we live is a part of the solar system. The parent body is the Sun, which is an ordinary star. The Sun appears to be close and hot and big to us simply because we're rather close to it. We're 93 million miles away and it's important to realise that the whole solar system is a microcosm of the universe. In fact the nearest star which is rather similar to our own Sun is so far away that the light from this star, which travels about 186,000 miles every second, takes four and a half years on its journey towards us.

Now, in order to appreciate the immensity of the cosmos, it may be useful to think in terms of its size if we were travelling with the speed of light (186,000 miles a second). If we set out on such a journey from Earth, we would pass the Moon in a second or so. We would pass the Sun in eight minutes and after about half an hour we would move out of the confines of our own solar system. We would then travel in interstellar space for four and a half years before we approached the nearest star. We could continue this journey for something like 100,000 years and every few years we would pass another star, which was in effect a sun. Now, in this local galaxy or Milky Way system in which we would be travelling are 100,000 million stars, ranged not uniformly but in a flattened disc and on Earth as a part of the solar system we are not at the centre of this disc as at one time believed, we're right out at one of the spiral arms, so far away from the central regions that it takes light about thirty thousand years to travel from the Sun to the centre of our galaxy, but these 100,000 million suns which make up our local Milky Way system are themselves an extremely small part of the universe to which we can penetrate today with our big telescopes. If we could imagine ourselves continuing on this fantastic journey with the speed of light, we would move out of this local Milky Way after about 100,000 years. We would then travel in interstellar space for something like two million years before we came to another great star system, the spiral nebular N31 and Andromeda and when we reached that we would find that the Great Spiral and Andromeda was, in fact, another system of stars, rather similar to our own Milky Way roughly having the same dimensions, about 100,000 light years across its major axis and containing ten thousand or 100,000 million suns. Today we know that the universe is made up of trillions of galaxies of stars like our own Milky Way and like the Andromeda nebula.

In fact, with our modern optical and radio telescopes, we can penetrate so far into space that we receive the radio waves and the light which had been travelling from these distant galaxies of stars for a few thousands of millions of years. They are, as we say, a few thousand million light years distant and within this great region of time and space the numbers of such galaxies, each of which contains ten thousand or a 100,000 million suns, are almost uncountable.

It is within this scope of the size of the cosmos to which we can penetrate with our instruments that one has to assess the possibility of whether life, as we know it on Earth, is unique in the entire universe.

Now there are two problems. First of all there is the astronomical problem, the second is a biological one. The astronomical problem can today be answered with a fair degree of certainty. It is whether these billions and trillions of stars which make up the galaxies in the universe are isolated stars or whether they are like the Sun and contain a number of planets around them, at least some of which might be capable of supporting some kind of life as we know it, and the situation there is that modern considerations lead us to believe that possibly a few per cent of the stars have planets around them like the Sun and that perhaps one per cent of these planetary systems have at least one body which has been in a stable configuration for something like 1,000 million years like Earth, which is roughly the time needed for organic evolution to occur as it has done on Earth, and therefore one can say from an astronomical point of view with a high degree of certainty there must be many trillions of planetary systems in the universe and extremely large numbers of these must have at least one of their planets in the kind of stable environment which is essential if life is to develop. This leaves, of course, the great biological problem as to whether the primeval [?] from which life emerged on Earth are in fact unique or whether they were widespread throughout the primeval dust from which the solar system and the stars and the galaxies were formed a few thousand million years ago. We have no answer to this question at the present time, although the investigations of the meteorites which come to us from outer space indicate the existence of [?] and it would, I think, be true to say that amongst many scientists there is a belief that life must be widespread in the universe. Of course, we live in an age where a definitive answer may soon be obtained to this great problem, because although nobody thinks that elsewhere in the solar system or any other planets of the solar system any forms of intelligent life have developed, nevertheless, if the primeval [?] from which we have emerged exist on the primordial dust then some signs of early development must exist, for example, in the atmospheres of Mars or Venus, and this means that the space probes which are now looming out into space and sampling the atmosphere of the planets and landing on their surfaces will, perhaps, within a decade or so telemeter back to Earth the vital information as to whether the primeval [?] living in fact exist on the primeval dust. If, as we do believe, the answer is in the affirmative, I think one must assume that life must have developed at many other places in the cosmos.

CARAS: If life does evolve and the time is protracted enough, may we assume that intelligent life is an inevitable result of evolution?

VOICE: Can we cut that bit and start on something else?

CARAS: Yes. If the life condition exists on a planet other than Earth and life in fact evolves there or occurs there can you envision it?

LOVELL: It is not known whether the ultimate and necessary consequences of any form of evolution must be intelligent life. It is presumed that the evolutionary trend along which life has taken on Earth does necessarily end up in some degrees of consciousness, but of course the extent of the development of consciousness with time is entirely unknown and one must presume that if organisms have survived to the state which Earth organisms have survived, then at least on some of the planets belonging to some of the stars in the cosmos there must be organisms in extremely highly developed states of consciousness – to such an extent that we can have very little conception on Earth of the degree of intelligence and consciousness which might have been reached or, indeed, the relation of that kind of consciousness to which we understand ourselves.

CARAS: Do you believe there has been an effort to contact Earth by radio?

LOVELL: The question as to whether any efforts have been made to contact Earth by radio can't really be answered. Of course, a number of attempts have been made and are still being made by people largely with telescopes on Earth to contact civilisations, but any claims to success have never been substantiated. Indeed, the problem is an immense one and the exploration of the total radio spectrum that limits the sensitivity necessary is a task which could not conceivably be carried out without an organisation of the sort on a global scale. It, of course, could be argued that the degree to which Earth's civilisation has developed is a very contrary phase. After all, the problems of radio communication or the possibilities of radio communication have only been realised on Earth during the last fifty years and we are considering the possibility of there being civilisations in the universe which have existed in a degree of consciousness for many thousands or tens of thousands or even 100,000 years longer than Earth civilisation and their interest and techniques would therefore have reached a degree which it is impossible for us to conceive. It may be said that they would know all about us and would regard as futile the possibility of communicating with us or would have given up the attempts many hundreds of years ago when they found we were not in a fit state to respond. The positive answer to the question cannot be given. It is only possible to speculate along the lines that other civilisations who would have the capability of communication may rapidly have achieved such a state of excellence that they are no longer interested in our own civilisation on Earth.

CARAS: Is there any reason to believe that Earth may have been visited at any time in the past?

LOVELL: As far as I know there is no reason to believe that there has been any visitation of Earth and I think it can be taken with a fair degree of certainty. Again, the same kind of arguments apply to a civilisation which achieves a necessary level of technology to carry out visits to other intelligent beings, remembering that in our case it would require a journey of five or ten years with the speed of light before we could reach another such civilisation, but perhaps by the time we have achieved this technological capability, we would have lost interest in the chance of acquiring contacts with other civilisations.

CARAS: Do you conceive other civilisations achieving such sophistication that they could attack life on our planet, without having any curiosity to investigate it or attempt to contact it?

LOVELL: I would have thought so. I think that if one reflects on the technological progress which is being made, even in the last twenty years on Earth, then it has an atmosphere of great drama about it and even without present facilities one has only to think towards the end of the century to see that the kind of techniques which might have developed make it impossible to conceive the potential which the civilisation might have and the state of its knowledge. This development might imply that during the next fifty years our intelligence will reach such a high state that our knowledge of the universe will become so great that we no longer have the necessary interest in carrying out these kinds of experiments as far as other civilised beings are concerned.

CARAS: Since our Sun, our star is not necessarily like all stars and therefore our Planet is not a very old planet is it possible that if intelligent life has come about elsewhere it might be even millions of years more advanced than our own?

LOVELL: That is possible. We are not all that young. After all the age of the Earth is 4½ thousand million years. The so-called age of the universe is about 10, 000 million years and it is unlikely that the stable planetary systems which are capable of supporting life can be tremendously older than our own Earth, factors of perhaps 50 per cent more, but of course we're talking in terms of thousands of millions of years and the question as to whether there might be a civilisation with another million of years of development can be conceived quite easily. The answer is that there may well be. If organic development has occurred elsewhere in the universe then it is very reasonable to presume that some of these developments may have been going on, not for a period of thousands of years longer than that on Earth, but for a period of a million or so years longer.

CARAS: You said it was difficult to imagine what it might be like but could you possibly speculate on that what might it be like if our civilisation were not ten thousand years old but one million and ten thousand years old? Would we [?] to our present biological imitations?

LOVELL: I think this is impossible to speculate on usefully on a scientific basis because of the rapidity of the advance of science. One has only to place oneself in the position of being asked this question in the year 1900 to speculate on the state of our civilisation, the state of its technology in the year 1950, 1960 or 1970, and this would be a matter of extreme difficulty. It would be guesswork. It would be regarded as science fiction and to be asked to speculate over such an extremely long period I think cannot be done from a scientific basis. One can speculate over a period of ten, twenty or even fifty years on the knowledge of our present scientific attainments by making certain assumptions about the stability of the civilisation but over longer periods I don't think the speculation is fruitful. I would say, however, that apart from the relevant immaterial sense, it is my opinion that the question which would have to be considered most carefully would be the possibilities of changes to the human being, artificial changes to the human being, particularly the human brain. The possibilities there are transfer of memory, for example, which would raise possibilities which at the present time are not only rather frightening but almost inconceivable as far as their consequences are concerned.

CARAS: Do you feel there's justification of our present sphere of knowledge to continue a conscientious effort to search the universe for radio signals and to propagate signals that might be received elsewhere and answered?

LOVELL: I think this will continue to be done to a certain extent but as a scientist I would not really believe in carrying out this task unless it could be done in a scientific manner and the problem is one of such magnitude that it is no good playing with it in the sense that, well, we wouldn't ask one set of people with large radio telescopes to search a certain part of the sky or a few of the nearby stars. This would, in my opinion, have to be done by a co-ordinated worldwide effort using almost all the radio telescopes and techniques which are available on Earth and at the present time there is so much astronomy, so much fundamentally we still have to find out about the universe that I think that it would represent an unjustifiable diversion of our efforts.

CARAS: What are your views, sir, about the origin of life on Earth?

VOICE: Can we cut that, please?

LOVELL: I can answer you straight away, you must ask the power on that [?].

THIRD PART / TAKE ONE

CARAS: Do you think our [?] are religions or intellect prepared for [?] contact due to come within the next few decades? What do you think the likelihood will be of severe cultural shock?

LOVELL: I think that the problem of the impact of successful communication with [?] intelligence if it ever came to that would of course depend on the nature of this contact. I think the very fact that it could be demonstrated that intelligence of our type existed elsewhere in the universe would obviously have profound philosophical and theological repercussions, but as far as our community generally is concerned the implication would, of course, depend on the nature of contact, whether it was a benign contact or an aggressive contact, and it is very difficult to speculate unless one makes some assumption about that. I think the interesting thing on which one might speculate is if the contact happened to be made with some civilisation which was in advance of ours but not so dramatically in advance that it had no concept of our problems or the perils of our own age. For example, if we could imagine ourselves making contact with a civilisation which was say fifty years, or a hundred or even a few hundred years ahead of our own then this could be of tremendous importance because we might be able to establish a type of communication in our own idiom and we might receive extremely valuable information which might guide Earth's civilisation over the immediate day-to-day perils which it faces, because it is also, I think, the natural assumption that many extraterrestrial developments must have gone through the problems of the atomic age with all its dangers. They would either extinguish themselves or the very fact that we were communicating with them would imply that they had survived and there might be some extremely valuable guidance to be given about the developments to be pursued on Earth which would help us in our future. You see, we happen to be passing into a unique phase of development on Earth where it is no longer possible even for the richest countries to support all the scientific developments which it is possible to carry out. In this our future is going to be sharply distinguished from our past in which the scientific work has been supported whenever it has been regarded as reasonable. This will no longer apply. In ten or twenty years there will be many areas of possible scientific development on Earth which simply will not be followed because the economy of the Earth will not allow it. Now, some kind of guidance from the civilisation which has passed through this phase and has survived it by fifty or a hundred or two hundred years could be of immense importance and this I think is a sort of optimistic outlook which one would like to have as regards the impact of successful contact with some other civilisation.

ADDENDUM: Sir Bernard Lovell, 2003

Thirty-seven years have passed since this interview. Although there have been dramatic advances in our knowledge of the universe since that time I am surprised that my original responses to the questions posed in 1966 have not changed substantially. In detail I would like to make the following comments.

1 At the time of the original interview an important discovery was made at the Bell Laboratory, New Jersey, of the cosmic microwave background. This discovery settled the cosmological argument between the steady state and continuous creation theories. It is now generally accepted that the universe originated about 14 billion years ago and evolved from a hot dense condition. This so-called 'big bang' concept is the current wisdom, but the early stages of evolution continues to be a subject of widespread theoretical investigation.

2 Although our detailed knowledge of the structure of the universe has substantially increased, the broad description I gave in 1966 remains reasonable. The deep field investigation with the Hubble Space Telescope in recent years reveals many thousands of galaxies in deep space on a very small area of sky in which only a few foreground tars of the Milky Way can be seen. The evolution of the galaxies of many types remains an intense subject of investigation.

3 In the years since 1966 it has been revealed that our observations reveal only a few per cent of the content of the universe. The current mystery under intense theoretical and observational studies is that more than 90% of the universe consists of dark matter or dark energy.

4 Advances in the details which might change my opinion about the existence of life elsewhere in the universe have not changed my opinion that 'we do not know.' In favour are the discoveries of planetary systems around some nearby stars and the discovery of interstellar gas clouds where the formation of stars can be seen contain many organic molecules and probably amino acids from which life emerged. On the other hand, it remains the case that in our own solar system the circumstances which led to the evolution of the Earth as a habitable planet remain unique.

5 Above all, the problem of consciousness remains a major issue. Although so much has been discovered about the human brain the nature of consciousness is not understood. We do not know whether any type of organic evolution in the universe would inevitably be associated with consciousness, as we understand it applied to human beings on Earth.

6 The technological developments on Earth since 1966 (for example, computer technology and its affect on human life) underlines my 1966 comments about our total lack of understanding of the technological development in any interstellar community. It is impossible to speculate.

7 Attempts to receive signals from other interstellar communities have been intensified with entirely negative results. This means little one way or the other, for reasons given above that we have no knowledge of the technological state of any interstellar community. In any case, with the most modern equipment, the sensitivity is such that given such communities have similar technological ability, our range is limited to a relatively small number of the nearest stars in the Milky Way.

8 My 1966 comments on the likely effects of any such contact are the same as I would make today.

Margaret Mead

[Curator of Ethnology, the American Museum of Natural History]

SCENE ONE / TAKE ONE

CARAS: Doctor Mead, there has been a lot of published material lately – popular level material, particularly from Russia – describing ancient finds, cave drawings, cave carvings, rock carvings, that seemed to depict men with space helmets. In all your experience, is there any evidence at all, anywhere, that either a culture has been influenced by an extraterrestrial visitation or that there has been anything that even resembles an artefact?

MEAD: No, we don't think so. Of course, it has been popular in anthropology at times to think of one culture somewhere that invented practically everything. This was very popular in the 1920s – Elliott Smith and Perry traced everything to Egypt, all over the world.[1] Once you make that sort of an assumption which is not justified you could then say, of course, that somebody landed in Egypt and taught them everything, but we neither think that everything originated in one spot nor, at present, do we see any reason for believing that there was any extraterrestrial evidence or visitor or anything of the sort.

CARAS: There is a budding new science we are told, up at Harvard, called astro-archaeology that particularly has taken great interest in Stonehenge and it worked it out in a computer that the complexity of the Stonehenge configuration is such that it simply could not have been worked out unaided by a primitive culture, there must have been a culture somewhere influenced, or very much richer than we realise at the moment.

---

1 Professor G. Elliot Smith in *The Migrations of Early Culture* (1915) and other works argued that all civilisation began in Egypt and eventually spread out to the rest of the world. His cudgels were taken up by W. J. Perry in *The Growth of Civilisation* and subsequent books. The theory had much currency in the 1920s and 1930s but is now discredited by anthropologists and historian alike. The idea that civilisation began in one place and radiated out is known as diffusionism or, more correctly, heliocentric diffusionism.

MEAD: Well, there must have been a culture richer than we thought built Stonehenge, that's all one has to say. We know when every time that we have looked at a very primitive, what we thought was a very primitive culture, and looked at it carefully when we have had magnificent stone remains or something of the sort, we find that they did more complicated things than we thought they could. You see, Western European culture is suffering from an appalling swelled head and it has never been willing to recognise that anything was done anywhere before the Greeks. The world started with the Greeks and ended with us and, of course, this is not so and there have been magnificent achievements in mathematics and architecture and astronomy in other parts of the world. It is a very easy out to say, 'Well, you know, must have been people from outer space,' and that still leaves us superior to all the other people on Earth.

CARAS: On the subject of extraterrestrial life: do you believe in it? Most scientists today seem to in most disciplines.

MEAD: Well, I think there is a perfectly high probability. A high probability does not mean that it is there. I'm more interested in the fact that we are going to look for it than I am, in a sense, in whether it is there or not because my field of competence is looking at people on this Earth and it is going to make a great difference to people on this Earth to stop thinking of themselves as the unique, perfect creation, and realise that there may be others.

CARAS: At the beginning, I am afraid you didn't mention extraterrestrial life. I am sorry. I was hoping that you would, so I didn't stop you. Would you rephrase it and make reference to extraterrestrial life as there is no reference?

MEAD: Some people have claimed that the only way we can explain some of the high civilisations of the past is by extraterrestrial life having landed here at some point and taught everybody, everything in things like Stonehenge for instance and the complexity of the astronomical measurements that they have found when they looked at it. Now this relates to a lot of our other attempts – in the past – to think of only one bright civilisation somewhere and give it the responsibility for everything else. In the 1920s Elliott Smith and Perry had the theory that everything came from Egypt. Now, if you had accepted the idea that everything came from one spot, it is only one step to imagine that space travellers landed and taught it to the Egyptians. But nobody supports that position today. We have excavated one complicated culture after another. We know that there were great complexities before the beginning of our particular era in Greece, and we know that we should not think only in the sequence from the Greeks to us in considering what complicated things other people could do. It has been very comforting to think that we were the most superior people on the Earth and if you start civilisation with Greece and end with us, this catered to our sense of superiority.

CARAS: What about the existence of extraterrestrial life?

MEAD: I think that most of the astronomers today agree that if life developed on this planet that there would be many other conditions in the universe where life might also have developed. Now, whether it would have taken anything like the same course that it took on this planet, of course, is incredibly hard to predict and those of us who have studied the development of man, and especially the development of human civilisation, realise on how many small accidents this development depended. I have not much sympathy with the point of view that thinks that life must have developed into something like us on a great many different planets and, you know, some people also build a picture which says that we must be in the middle. On the normal probability there should be as many civilisations older than we are, as there are younger. I think that is a piece of nonsense reasoning when you are dealing with history.

CARAS: On this subject of age, if I understand correctly, would there not be likelihoods of civilisations very much older than our own?

MEAD: There is a perfectly good likelihood, of course. There may be older civilisations – if there is life that is anything like ours. When we talk about civilisation we are talking about creatures that depend upon learning and invention instead of depending upon their genes. We are talking about creatures that could not communicate with each other properly at all if they did not learn a language, and a language is a human cultural invention. Now, this is very specialised to human beings. It is the reason we have been able to accumulate from past generations and each individual is able to make a contribution to our culture which does not depend on being transmitted by genes but can be transmitted by books, and today by tape and film, and all of these other things. Now, we would have to assume if we talk about civilisations on other planets – that on other planets creatures evolved who ceased to depend upon genetic transmission of their particular capabilities, and instead depended on learning and stored learning so that they had something comparable to a culture. So there are a great many ifs in this whole picture that are much more complicated than most of the speculators admit.

CARAS: Can you envision life arising on any planet? Any kind of life? And not undergoing an evolutionary sequence of some kind? Could it be static?

MEAD: If one thinks about life as we think of life, we think of life as something that changes. We think of a process, So that it is difficult to envisage life on a planet that would be eternally static in which there would be no change but it is, of course, perfectly possible to envisage or imagine very slow change, very different rates of change from the rates that we are dealing with on this planet.

CARAS: If an evolutionary sequence was launched, or was triggered, would intelligence be the inevitable result, given enough time?

MEAD: If we want to deal with the problem of whether creatures on another planet would be intelligent in the sense that human beings are intelligent, one has to assume that implicit in the beginning of life in some way were the stages that happened to have produced us. This bars out the possibility of accident. It assumes some unidirectional factor in evolution that I don't think we have any justification for either believing or disbelieving today. We don't know enough.

SCENE TWO / TAKE ONE

CARAS: Doctor, civilisations the world over down through the years have withstood terrible impacts, terrible shocks, but what would the cultural shock be to various types of civilisation, from sophisticated to crude – if crude is a moderate word – once it were established that there was extraterrestrial life, particularly intelligent life if that should be, and if there should be radio contact? What kind of reactions might we expect?

SCENE THREE / TAKE ONE

MEAD: If it ever should be established that there is extraterrestrial life somewhere in this universe that we can communicate with, even if we could only communicate by radio contact as many people feel is the most likely today, many people feel, of course, this would be a dreadful shock to human beings on this planet. I am not so sure it would be. If one looks at the way they have adjusted to the whole idea of satellites in space, how rapidly satellites became just everyday events, so that the children today are absolutely accustomed to satellites, and it is only the people over forty that feel that satellites are something strange or unusual. We would have a whole group of older people who would probably have fits and would find it even difficult to remember for any length of time that there were people – creatures – somewhere in the universe that we communicate with. But man's capacity to adjust to new situations is being steadily accelerated now. We have made studies in this country, just at the time that Sputnik went up, how people responded to it. Before Sputnik about thirty per cent of the American people could not spell 'satellite' and most people did not believe that it was possible and many scientists were saying that it was hundreds of years off before we would do the things we have done in the last ten. So, for every marvel, and it is the marvels of course that will make it possible for us possibly to explore outer space, for every marvel that occurs human beings adjust a little further to the expectation of the next marvel, which in turn becomes less marvellous.

CARAS: What about philosophies? Is there room in the philosophies currently with force in our lives that allow for extraterrestrial life?

MEAD: Oh, yes! We are behaving much better than we did at the time of Galileo when we really could not allow for him, but it was over I think about seven years ago that when one of the international astronautical societies met, the Pope stood up on his golden throne and blessed exploration into God's universe[2] and I am sure a good many religious discussions are already going on as to how we are going to incorporate these creatures into our religious systems. We had a great deal of difficulty incorporating American Indians, you know, and we have not really successfully incorporated the human race yet. We are still working at that and the process that we used to incorporate the whole human race, whether they are great giants in Africa or tiny, little pygmies somewhere on a small island like the Andaman Islands, the same efforts that we had to make to face the fact that we were all members of one species on this planet will simply have to be made on a larger scale if there are other creatures that we have to include, and this has been going very fast. I found when I made a speech in West Berlin a few years ago, and ended my speech with a great, inclusive statement about all human beings on this Earth, brown men, yellow men, red men, pink men, black men, and it always sounded very inclusive, but it didn't sound inclusive any longer. It sounded narrow and parochial... provincial.

SCENE FOUR / TAKE ONE

CARAS: These contactees who have been to Venus and so forth somehow always described extraterrestrials as wearing long white robes, having long white beards and aquiline noses. What does this tell us?

---

2 Dr. Mead is thinking of Pope John XXIII's address to the Pontifical Academy of Sciences in Rome in 1960.

MEAD: If we look at the different sorts of images that people have built up when they have claimed to have made trips to Venus or to Mars and have images of what the people are like, they are of course a variety, they are religious images that fit very well into some of our pictures of the Old Testament prophets, or the Apostles, so that people used the religious signature they have to build up their pictures of other worlds and then, of course, there are the incredible imaginers of very, very different forms of life and then there are the people who assume that the inhabitants of other planets look exactly like us and you would not be able to tell the difference. I think this is probably the most foolish bit of all.

CARAS: In the subject again of philosophy. There is no great conflict in our philosophies that would deny us belief in extraterrestrial life?

MEAD: Man has been struggling with the attempt to include other members of the human race in his select society for a very long time and if one looks at the development of philosophical approaches to man – by man – you have to go back to very primitive tribes who did not usually admit that the people on the next hill were fully human, and we have been expanding our definition of humanity right through history. What we really mean by great religion is the religion that has been able to include members of other races with different customs and different languages within the category of creation in which we have placed ourselves. It is only in this century that we have fully known that human beings did belong to one species, that it was not just an act of religious imagination to speak of us as 'all brothers' and so right through history you have had a lot of difficulty and if you go back to the discussions when the Americas were discovered, were American Indians to be treated as fully human? We have had the same sort of discussion about Africans, about South Sea Islanders, and great religious struggles to phrase the whole of the human race on Earth as one race. Now the same problems would confront us if we have to include in our thinking creatures on other worlds. It will be very good for us, it will stretch our imaginations and, of course, there will be all the same conflicts. Are they likely to be very superior to us? That will make us very frightened because we have been very unkind to people whom we thought inferior and we will project this on our image of extraterrestrial life. On the other hand, if we think of them as very inferior, that may frighten us a little too because that will reactivate the sort of conflicts that we have been dealing with and trying to get rid of, imperialism and colonialism. So, any knowledge of extraterrestrial life will mean a lot of rethinking of the universe and man's place in the universe. I think it is very good for man to have to rethink his place in the universe.

CARAS: There is nothing about intelligence for us, say, that is the learning and the process that says that morphology would have to be the same as ours? Of course, what happens if we find something whose overall morphology is totally different from ours but who is as intelligent as we are, or potentially as intelligent if younger, or more intelligent, more sophisticated, if older? We will call them man?

MEAD: There is, of course, the whole question of how we would be able to relate if we find extraterrestrial life that is as intelligent or more intelligent that we are but morphologically extremely different. If they have senses, they might be very different senses from the ones that we think of so easily. Their methods of communication might be entirely different. Their methods of assimilating and processing information might be enormously different as for instance Hoyle mentioned in *The Black Cloud*.[3] His picture of the cloud's form of thinking being something that terrified the physicist so that he went all to pieces and died from the effects. Now, I think most people today have very narrow notions of the possibilities of the human brain, and they tend to think that the high processes of the human brain are very special and limited in quality. If we continue to develop our understanding of different kinds of communications and use different kinds of communication models we ourselves ought to be able to think systematically about creatures that are very different from ourselves and work out methods of communication. If, as some people would suggest, there are creatures far more advanced in forms of communication and information processing than we are, then it would be up to them perhaps to develop first a means of communicating with us.

CARAS: All the hysteria and all the goings on about UFOs. Even if there are UFOs there has certainly been an awful lot of nonsense as well. Is this some kind of conditioning that people are going through to prepare themselves for the fact of it? Are they living with a fiction enlarged part so that one day they can live with the fact? What is the meaning of this in our culture?

MEAD: It was suggested by Jung a few years ago that all the appearances of flying saucers, all the stories about flying saucers and the claims about them, or as they are now called, UFOs, was part of the preparation of the spirit of the age of the folk mind for satellites and that people were going through a kind of pre-figurative, prophetic state before the real events occurred and interest

---

3 See Footnote 1 in the I. J. Good interview, *supra*.

in UFOs subsequently went way down. What happens is that if some new space activity, very new and startling, the first time that a man walked into space occurs, you get an outburst of UFOs, another big outburst and then it dies down again and people begin paying attention to the reality rather than to myth and the dream and then with the new extraordinary effect you get another outburst. I am pretty sure that if we succeed in landing on the Moon and staying there for any length of time, we will have an enormous attack of extraterrestrial travellers coming back with accounts of what they ate for dinner, almost immediately.

CARAS: A little projection to the future. The date could be argued but let us say, roughly, that if stretched to the breaking point civilisation as we know it in the Western world was, say, ten thousand years old, an arbitrary figure. If there are extraterrestrials of much longer spans in time than we have had, very much more sophisticated, if they are intelligent, what might they be like? What might their capabilities be? And what would their philosophies be if they had civilisations four million and ten thousand years old instead of ten thousand years old?

MEAD: Of course, it is very hard for us to think what very old civilisations, much older than ours, would be on other planets in the universe. The best way to try to do it, of course, is to think oneself back beyond ten thousand years or so that we give our civilisation and try to project forward. This is where anthropology is very useful because when you work with a people who are living way before civilisation, who had no script and had no metals, and had no form of government like ours and had no religious beliefs except in a few wandering spirits and ghosts around under their own trees, and then you try to think from there forward to where we are now, you get some notion of the orders of change that could take place. Now, I think the major problem in trying to move from say, civilisations that have had ten thousand, twelve thousand years to move and develop, to go from them to a notion say of four million years is that we have not any idea really about rates of change. In our own society change has been accelerating and accelerating and accelerating, so that if you use a clock you speak of man as having come in way after the eleventh hour and of modern civilisation as only being a few seconds, as against the whole history of life on this planet. Now, we do not know whether, as civilisation becomes more complicated, it will become more stable. This is a possibility, just as human beings are actually very stable because most mutations are lethal in creatures that are so complicated. Now, it is possible that we could build a civilisation that was still built the way we built it, that is by learning and the transmission of learned material that would, nevertheless, be so complicated that you would not dare monkey with it. We have one example of a civilisation, not a civilisation in the sense of 'high civilisation,' but of a culture of this sort among the Eskimo which has been incredibly stable for a very long period and the language is spread over a very wide area because they were living adjusted to such a dangerous…

SCENE FIVE / TAKE ONE

CARAS: Doctor Mead, Dr Harlow Shapley was telling me just yesterday at Harvard that twenty-five years ago if a graduate student said to him that he wanted to question or study the origin of life or to discuss or investigate or do a paper on the possibilities or probabilities of extraterrestrial life, he would have told him to leave the room and close the door quickly as he left. Told him to go and look at some stars, record some luminosities and do something useful. He confesses this and he says today of course is totally different. Twenty-five years is a rather short time for such a change, isn't it?

MEAD: Of course, when we look at what has happened in the last twenty-five or thirty years and how problems that would never have been respectable and that were regarded as science fiction. The sort of things that H. G. Wells would write about have become respectable objects of study. Our people are working on them, the origin of life and working on the problem of communication with possible residents of other planets. We are inclined to think that this has moved terribly fast and some people, specially people who grew up fifty years ago and I'm sure people who grew up seventy-five years ago, are a little bit terrified by this rate of change. To others of us, it is very congenial and if you have looked at the whole history of man you realise that the cumulative nature of man's technical knowledge means that the faster you go, the faster you go. Now, if we were depending on our genes instead of on cumulative knowledge we would not have change at this rate and we would be like the ants, many of which have not changed for incredible lengths of time, because their stability is due to their morphology rather than to human culture as ours is and our culture is not nearly as stable and it will accelerate rapidly and we can expect tremendous acceleration in the next twenty-five years. I am sure, and on the whole almost every scientist has been too conservative in his estimates, they are still being too conservative, they are still saying we won't do something for twenty years or thirty years, and we do in the next month.

CARAS: Machine intelligence. The intelligence explosion. What change is this going to bring about in our lives?

MEAD: Now, of course, the new information explosion, the new cybernetic revolution in general, is going to be a change that is probably greater than the Industrial Revolution and people who have studied the suffering and misery that followed the Industrial Revolution are inclined to think that we have to go through the same kind of suffering today. There are many people who talk about automation as if all it was going to do was to throw people out of work and put them on the poor rolls or on relief when, actually, automation is going to make it possible for us to do things we never did before and the information explosion is going to make it possible for us to work at a quite different rate. We are going to have to invent ways of using it, of course, and one has the picture of every student with a little computer sitting on his desk and being able to ask the greatest stores of information in the world for the thing he needs immediately. The whole of human knowledge will cease being handled in the linear fashion and will be able to work simultaneously. This is very congenial to the human mind. Great human minds have always worked simultaneously. They have always thought about twenty things at once at different levels and with different senses so that I think that the new machine intelligence, the new automated world, is not only going to make it possible for the first time to feed people, to care for them, to protect them, but also it is going to release tremendous intellectual potentialities that have never been used before.

CARAS: The prophets of doom that take extraterrestrial life and make them inimical to our way of life, they will take a computer and make it a monster that is going to take over and make man obsolete. So, are these the same people who said we should not fly because if God wanted us to fly we would have wings? What is the nature of this kind of person?

MEAD: We have always had with us people who were afraid of change, of course.

People who said that man would never fly or man should never fly because if God had meant us to fly he would have given us wings. They objected, of course. They objected first to the motor car and I am sure if we had the records, of people that would have said that a sail was something God never meant us to have either. If things are far enough back then we reassign them to God's approval and almost every invention before the birth of Christ is regarded as good by the people who think that almost every invention since is bad. You have always had in human societies – though they are worse in some societies than in others – people who feel that any change is dangerous. They are rigid people. They have grown up to depend on a very, very narrow set of habits and ideas. They are frightened of change and so they turn any kind of change into a sin, an impiety against God's will, or they make it devilish and they think computers are going to take over and manage us or even, I have heard people say, 'Maybe the world's being run by a computer now.' That is a devastating thought. We have always had those people, 'The Lord meant us to stay on the ground,' and then, 'The Lord meant us to stay at least inside our own atmosphere,' and when we are communicating with other galaxies we will have people who will say that six galaxies are enough, or something of this sort and if the Lord meant us to go to seven, He would have arranged it.

CARAS: Is there anything you want to add to any of the subjects we have discussed? Machine intelligence? Extraterrestrial life? Cultural shock?

MEAD: Of course, the most important thing, I think, in preparing for these changes is to have new methods of education and to realise that we are going to have to apply those new methods not only for our own children but for all the people on this Earth. We can't think of people who were primitive yesterday, in the centre of one of the southern continents or in the islands of the sea, we can't think of them as our ancestors, we have to think of them as our contemporaries coming into the same world we are coming in to and needing the kind of education that will fit them to use computers without having gone through all the stages of early script and early mathematics that we went through. Human beings never have to go through the same stage. If one group of human beings have learned how to do things, another group can learn. So, at the same time that we are developing new methods of education for our own children we have to be ready to have those available for people in every part of this globe, realising that whether there is extraterrestrial life or not, we are one human species on this Earth and we all have the same potentialities, and we are all facing this new world together.

[This interview is reproduced courtesy of the Institute for Cultural Studies, Inc., New York.]

SK with Kier Dullea inside the 'Hal's brain' set

Marvin Minsky

[MIT Artificial Intelligence Project]

SCENE ONE / TAKE ONE

CARAS: Doctor, what do you see as the ultimate development of computer intelligence, machine intelligence?

MINSKY: There is no reason to suppose that there is any definite limit to the intelligence that may be achieved by machines.

CARAS: What progress do you see being made in the next ten years? Sort of timetable, if you will. Extrapolate a little bit.

MINSKY: To predict what will happen in ten years in the development of machine intelligence is to predict how hard people will work. It is very hard to be sure of anything in this field. The field is only ten years old.

CARAS: What do you think will be the first great changes in our environment or way of life as a result of any machine intelligence at our disposal?

MINSKY: I think there are two directions in which machine intelligence will affect humanity very deeply. In one direction are the enormous increases in production, productivity – the replacement of work by leisure, the availability of luxuries.

The second direction is the intellectual awakening that we will encounter by meeting aliens because I think the first aliens that we will meet are not aliens from other galaxies but aliens of our own creation: intelligent beings whose nature is profoundly different from our own, yet which because of the way we have designed them have similar interests.

CARAS: Such evidences created by man or by man's machines, of super-intelligence, will they be electronic or biological machines?

MINSKY: When we finally develop a highly intelligent machine it will be in the same period that we have highly developed computers and presumably it will take the same general form with smaller and smaller parts, not resembling biological components very much, I think, but resembling them perhaps in size, much greater speed, much greater memory capacity and physically rather small.

CARAS: Will the time ever come when machines of this nature will have emotional problems? Are we going to end up with neurotic machines that will be intelligent enough to put each other on the couch and analyse each other or themselves?

MINSKY: Intelligent machines will be subject to kinds of behaviour that we might call disordered. These aberrations, neuroses or whatever we will come to call them will not necessarily resemble the kind of disturbances that humans are subject to but they will behave irrationally. Perhaps in the evolution of the thinking process of one of these machines it will use some technique which is good and useful on easy problems but which becomes more and more cumbersome on difficult problems. When that happens the machine may not be able to discard it very easily. It may be woven into the web of all its other techniques in the course of its intellectual development. And to correct it may involve the necessity of regressing to an earlier stage in the machine's period and that kind of behaviour we might call neurotic or cancelling, disturbing its behaviour, taking back things it has said and things like that. It won't be easy to remove these.

CARAS: Do you think the time will ever come when machines will have such sensibilities that the dividing line between man and machine will become fuzzy or hazy?

MINSKY: I think the time will come – I don't know whether the year 2001 will be that time – when the intelligence of machines is comparable to humans and that the problem of integrating them into our society will have to be faced. The problem of turning off the computer today is not a moral problem, it's an economic one. The problem of turning off a machine as intelligent as you with perhaps similar interests or alien interests whose value you are not certain about will be a moral question and the transition between these phases will be fuzzy. We will have to develop new ways of thinking when we share the planet with other beings of comparable complexity.

CARAS: If we are going to make a machine that is as complex as man or as intelligent as man then surely we are going to make one that is even more so?

MINSKY: To build a machine as intelligent as man is beyond our ability today. On the other hand, the intelligence of the machines that we have built is growing very rapidly by any kind of standard. But we do not know how the course of this development will run but as the machines approach human intelligence there is no reason to suppose that that will be a limiting growth. They may pass us by very rapidly and not stop for a great length beyond that. There is no reason to suppose that human intelligence is a particular terminal point or marked region of possible intellectual behaviour. It is probably limited by the physical size of our brain at present and certain things in its structure, probably nothing fundamental about it.

CARAS: Morphology. You were working in an area where you were breeding devices and mechanisms that can do intricate manipulations. Scientists have been working in the area of these super-intelligent machines. Will these ever be enclosed within a single morphological unit which will have a machine that can walk, can perhaps go with voice input, output, therefore talk, can manipulate with its artificial hands and can think? Will we have a robot as such in this morphological shape or is intelligence always going to be enclosed in computers that fill rooms and rooms and rooms and are worked by radio to a simple mechanical device?

MINSKY: The physical form of intelligent machines is not really very highly constrained by anything. It is possible that we will build machines that are the shape of man with hands the size and comparable dexterity as men and with the brain somewhere in that body. On the other hand, there would be no difficulty. One form the machines may take if the computer required for intelligent behaviour is still very large is to have the computer in some remote location. One may build a robot with hands, eyes, arms, legs – all the things that a man finds convenient for manipulating the world and have the central intelligence somewhere else connected by radio link or some other form of communication. There is no particular reason to put the brain in the body if the brain has to be very big. If it is small then you can put it anywhere you want. The body could be on the Moon, the brain could be on Earth, provided that one could cope with the delays due to the speed of light.

CARAS: Might it not increase man's uneasiness for tomorrow – the question we raised before – the more machine looks like man, the more of a problem it becomes for man to deal with it?

MINSKY: I think people will resist the construction of robots in their own form for rational or irrational reasons. One does not feel very threatened by the present computers. They are large. They have little lights that blink. They don't move. As the machines become more humanoid, if for industrial or for other reasons this comes about, I think people would be more disturbed and this is healthy. I think that people will have to face in the near future the problem of other beings which are not human, of sharing their life and planet.

SCENE TWO / TAKE ONE

CARAS: When are we going to get a breakthrough on voice input? How close are we?

MINSKY: Communicating with the computer in most applications today is done by typing, on teletype or a typewriter or something of that sort. It is possible now to talk to a computer in a very limited sense. There are devices which can interpret a hundred different words spoken by a person who is trained to use the device quite reliably. I think in five years any person would be able to speak to one of these devices without training. In ten years we should have reasonably good input for a large vocabulary. Certainly by 2001 one should be able to talk to computers in an ordinary tone of voice.

CARAS: When will they answer us, though? What about voice output?

MINSKY: To have the machine return its answer in English could be done today. We have programs that can produce a human voice-like sound although it sounds rather hollow and monotonous today and we have programs which can transform the results of their calculation into English in a printed form. So putting these two things together could be done yesterday. By 2001 the quality of the spoken output should be very high indeed. It would be very easy to make computers which appear to converse with you. The question of how full its understanding would be is open. It depends how much progress is made by that time.

CARAS: What are the advantages to having voice input? How about voice output? Is there an advantage to it?

MINSKY: I think voice output is quite convenient because it means that the operator or the user of the computer doesn't have to remain at some station keeping his eyes busy. As for voice input to a computer, there are serious difficulties with that, not in the technical problem of making the computer understand the spoken voice but in the problem of expressing yourself accurately enough; saying just what you want to say in the ordinary language. For most purposes that computers are used for, English or natural language is not a very good vehicle of communication. One wants to be more precise than that.

CARAS: Science fiction writers for a long time of course have been predicting the problems we will have with computers. They are going to take over and there has been an awful lot of nonsense written and an awful lot of good fiction written as well. What are the elements of danger? Are there any possible grounds for concern that man could be made obsolete or that machines can take over or that machines can do diabolical things?

MINSKY: Intelligent machines can certainly take over a great deal of our work. It is not obvious that many people would object to this and perhaps the danger to having the machines take over too much is that most people will welcome this. As for that, I think you can say that surely more than half of us are already not working in any real sense; most people aren't doing things that are absolutely necessary for their survival or for mankind's survival and I don't believe the problem of what to do if the machines do all our work is as serious as many people like to think. Some people don't really like to do work that they don't want to do very much. The problem of the machine taking over in a dictatorial sense of running the country and making people do what they don't want to do is a real problem. One must not delegate too much responsibility to another person or another machine or to anything else unless one is prepared to take the consequences, and one must consider very seriously the question of letting other people or other machines make decisions for you.

There is nothing special about the intelligent machine. There is certainly no reason to expect it to be more hostile or inimical than another person – perhaps less so since you have an appreciable amount of control in the way it is built and the way it is going to operate, so that there is no special danger here, I think. With respect to the machine that is super-intelligent that is far more intelligent than a human, if we create those then one cannot assess this. One doesn't know and one may have great difficulty in deciding whether to turn over any authority to it. One doesn't know what its goals are. One doesn't know what it will think is best for us and perhaps one may not understand its workings well enough, once it is more intelligent than we are, to control it. Just because we will be able to build machines that are intelligent doesn't mean that we will understand them very well. I'm afraid that the course of development of these machines will be a kind of evolutionary engineering process as in many other things. We will learn how to make them better, how to improve their ability to reason, to handle large concepts, large numbers of facts, do logical deductions, behave more and more rationally, but we may do this by cuts and tries and making small modifications until the first highly intelligent machines we develop will probably be great jumbles of complicated techniques. We may not understand them very well. They may work reliably for some time and then behave in a most irrational manner because of some imbalance in the way they assign importance to different factors involved in whatever they're trying to do.

CARAS: If we don't understand them would they understand us?

MINSKY: I don't see any reason to suppose that intelligent machines will be particularly good at understanding how human beings reason. The way that human beings think is evidently very, very complicated, quite disorderly in some respects. There is no reason to suppose that a machine, unless it is very much more intelligent than we are, will understand it very much better. As for the super-intelligent machine, who can tell? It may be able to understand a person at a glance, predict exactly what he'll do in innumerably complicated circumstances.

CARAS: How many years is this super-intelligent machine away?

MINSKY: Well, things are moving along quite rapidly. I think that in thirty years we should have machines whose intelligence is comparable to man's. I can't predict what kind of difficulties we will encounter in the search for this. It may be that at that point there will be no particular difficulty in making it much more intelligent. I don't even know what it means to be very much more intelligent than a man except in small ways, like having better memory, better ability to visualise geometrical situations, better mathematical capacity and so forth. But I don't see any reason why the machine should stop at that point. Why should we stop if we can achieve humanoid intelligence? We ought to be able to go very much faster. If we achieve human intelligence we ought to be able to go very much further too.

CARAS: What about motivation? Will a machine ever be motivated in the sense beyond doing what it is told to do?

MINSKY: We already know from the kind of work that is being done in cybernetics today that the motivation problem for a machine is not a particularly difficult one. It is quite thoroughly under-stood. It is easy to build a machine that has goals. It is easy to build a machine that has relatively abstract goals providing that we can formulate them, or give the machine enough clear-cut examples of what such goals are. There isn't any difficulty in making a machine, at least in principle, which will pursue very complicated and abstract goals that are the kind that we respect in other people.

CARAS: When will the first machine be conscious?

MINSKY: I think there is no line between conscious and unconscious behaviour. As a machine becomes more complicated, even in some of the machines we have today, the machine is itself dealing with statements in its own workings about the kind of goals it is trying to achieve. In other words, the machine is not just solving a problem on the simplest level. Some of our computers' programs are trying to solve problems about which of several goals are more important at the moment and I believe that one has a rudimentary form of consciousness in this. The greater the extent to which the machine is making calculations concerned with its own goals, the greater the sense in which the machine could be considered conscious. I think that when we get a machine as intelligent as humans I feel quite sure that it will behave in every way as though it is conscious.

• • •

SCENE FOUR / TAKE ONE

CARAS: The project you are working on here now is called artificial intelligence. When will machine intelligence cease to be artificial, or by definition will it always be?

MINSKY: When we talk about artificial intelligence that's just a name. If a machine is intelligent it doesn't really matter how it came about. One wouldn't call it artificial if it really solves problems for you in the same way that another person would. So that term doesn't reflect anything except the origins of such kind of work.

CARAS: How soon will we be able to entrust machines with jobs where failure would be extremely hazardous to human beings? For instance, when will the first passenger aircraft take off with a completely robot crew? In effect, will machines ever be that reliable? When will we entrust these things to them?

MINSKY: Computers are very reliable in situations today where we understand the kind of things that may happen. I think the computers of today could fly the airplanes of today with greater reliability than pilots can. There have been proposals to make completely automatically operated aircraft systems. I think the trouble is that people are reluctant to venture into the transition period where accidents will happen as the new, complicated system is being worked out. It is bound to have some troubles and we would like to put off, postpone, not face the dangers involved in transferring over to that mode of operation. I don't see that we should be as reluctant to do so because we could certainly begin by doing this with freight aircraft. There is no reason why any human lives should be risked in the experiment. Our spacecraft today are flown by computers.

I think the existence of astronauts in the spacecraft is an expensive luxury. If we could replace the equipment necessary to keep the astronaut alive by mechanical repair equipment adequate to maintain and adjust and make changes in the spacecraft in emergencies, we would come out ahead. Another statement is that if, as for intergalactic travel, I think, that if we are visited by aliens here I would be very much more surprised if the visitors are biological organisms. I think the most likely kind of alien visitor will have our machines, intelligent machines created by other races, on other planets. Because a machine could be built that could travel at less than the speed of light, for millions of years, maintaining itself and, if we were visited by aliens, it is much more likely to be searching machines, intelligent machines that have millions of years to spend, rather than organisms that are too fragile to stand such trips.

CARAS: Do you think there is any possibility that we have been visited in the past?

MINSKY: It seems to me that any culture advanced and powerful enough to send visitors to Earth from these vast distances, would either do it in such a way as to make it obvious, or else would be able to conceal itself completely if that were its purpose. The intermediate stage of occasionally sighted ghostly visitors I think is silly. A culture advanced enough for interstellar travel would be advanced enough not to leave silly clues if it didn't want to and if it did then it seems that it would present itself in some sensible manner.

CARAS: How long do you think it will be before we get concrete evidence on radio waves, artefacts, or what-have-you, of an extraterrestrial?

MINSKY: That is something you should ask an astronomer about.

CARAS: If we encounter a civilisation that is millions of years more advanced than our own, which is a possibility, what might their capabilities and machine intelligence be? What might it be like to have our ten thousand years of civilisation, and our twenty minutes of science, which we have on this planet, plus four million years, let's say? What might their capabilities be like? At what scale could they operate?

MINSKY: I think if we encounter a culture that is millions of years more advanced than us, or even thousands of years more advanced, then they will probably have replaced their limited biological bodies by mechanisms of greater capacity, of greater intelligence, and longer lifetime. I think man himself in a thousand years may replace himself by machines. He would be foolish not to. We are too fragile, we are too feeble and there are so many ways in which an ambitious man would like to improve himself that I don't think he will be able to resist the temptation. A thousand years in our culture will transform us in a way that most sceptics think will take billions. Our present culture is only two hundred years old in the most material sense.

CARAS: Will replacing our own biology with a superior machine, a machine superior to our bodies, will this enable us to do far more remarkable things in space than we could ever hope to do as human beings?

MINSKY: I think if we are going to visit the planets we can do that as men at great trouble and expense. If we are going to visit other stars there are few of them that could be visited by human expedition. If we are going to visit other galaxies we will have to rebuild ourselves into a creature more suitable for such arduous trips. We will have to make a creature that could stand millions of years of inactivity and we'll do this by transforming ourselves into machines or at least building representatives of ourselves that meet our standards and can make the trip.

CARAS: A semantics question. Will this machine be sexually derived and then improved? Will this be genetically engineered or will this be something totally biological, this replacement for man?

MINSKY: If man is replaced by machines, if he replaces himself by rational decision by modifying himself, which may be more acceptable than simply replacing himself with something else, he will only do this if he is in a state of rational conviction that we cannot imagine today. If he chooses to give these machines some kind of sexual reproduction ability that would be very strange. It is not clear what function it would serve. If one had sex at all in machines perhaps you would have a great many sexes? The chromosome system uses two gametes, for historical reasons. Undoubtedly a better way to reproduce than human beings would be to have twenty-three sexes and take the best chromosome from each instead of a random combination from both. So, the exact nature of the best reproduction system that we could design is not clear today. A few people have thought about sex from the point of view of how it could be improved. What we call emotions in man and think of as something apart from intellect is a very difficult class of things to separate. When a man is confused or neurotic, when a man behaves in a childish manner, one says that perhaps he is emotionally disturbed. But we do not know. Perhaps he is deciding in an intellectual way that his ways of making decisions normally are not adequate for this situation, that he should back off to a period or to a state where he has other ways of making decisions. Perhaps sometimes what we call emotionally disturbed behaviour is a rational thing to do. It's very hard to say.

Ormond G. Mitchell

[Associate Professor of Anatomy, New York University College of Dentistry]

SCENE ONE / TAKE ONE

CARAS: Doctor Mitchell, do you believe there is any possibility in the foreseeable future of our contacting extraterrestrial life? Do you believe that it exists?

MITCHELL: I believe that the possibility of contacting extraterrestrial life is very, very good. I think that the people today who are just initiating studies of this kind indicate that there must be some life somewhere in the far reaches of space.

CARAS: Is there any scientific argument that you know of that would militate against the existence of extraterrestrial life on other planets?

MITCHELL: I don't think that as we know life that we can say that it doesn't exist beyond our universe – perhaps in the outer reaches of the universe. Within our own Sun area I don't think that we'll find perhaps more than fossilised life or the beginnings of life or the evidence of previous life. In other words, I don't think that on Mars or the Moon that we're going to find other than the evidence of the formation of our universe.

CARAS: What are your feelings, theories or thoughts about the origin of life on this planet or on any other? How did it arise?

MITCHELL: Well, life arose, I believe – well it's been documented and I think the evidence strongly favours the ideas that were expressed years ago by Oparin. Much more evidence has been given to these theories of natural origin of life by chance by Urey[1] and other workers. The primordial sea idea that as the Earth cooled and water gathered on the surfaces of it, that the

---

1 Harold Urey (1893-1981), the American chemist and Nobel laureate (1934), and Stanley Miller (1930-2007), at the University of Chicago in 1953 conducted an experiment that simulated the primitive atmosphere of the Earth and demonstrated that amino acids and other essential materials of life could be produced in this pre-biotic soup. See I. S. Shklovskii and Carl Sagan's *Intelligent Life in the Universe* (San Francisco: Holden–Day, 1966), pp. 227-45, for an account of this and subsequent experiments.

chemical combinations were such that chance alone played a great part in the formation of life.

CARAS: Do you give any credence at all to the theory held by some that it could have arrived here as spores in a cosmic dust? That its origin is not terrestrial at all?

MITCHELL: I wouldn't like to answer any questions in that area.

CARAS: When life arises as it did on this planet and as we can assume that it has on other planets somewhere, is it inevitable that it should undergo an evolutionary process or is it possible for life to be static?

MITCHELL: I think that from what we see in the cosmos itself in evolution, that in evolution nothing is static. Everything is evolved. I don't believe that life would exist in a static state. I think that to survive, the organisms must change things that we know as living material.

CARAS: If life then arises, an evolutionary process goes forward. Is intelligence inevitably at the end of an evolutionary train?

MITCHELL: I think perhaps this is a bit of a selfish attitude that people take since we are intelligent beings and are analysing life in this aspect. I don't think that it necessarily has to progress toward life that recognises itself, if this is the definition of intelligence that you wish to give it.

CARAS: Is there anything else in this general area you want to add, comment on?

MITCHELL: Well, only the point I think that's being made that in the next ten years we will see men landing on the Moon and gathering evidence of the evolution of our own universe and I think that these will be the areas that will give us evidence for life beyond our particular little tiny niche in the cosmos.

CARAS: You are using universe then in the modern sense, of universe as instead of galaxy and cosmos instead of… ?

MITCHELL: Yes. To point out that there's a lot beyond this little Sun of ours, billions and billions of suns…

CARAS: If a life form arises and has four million years or five million years longer to evolve than we have, would you care to speculate what might it be like? What levels of biological or intellectual achievement might it have to its credit?

MITCHELL: Do you mean four million years beyond the formation of man?

SCENE TWO / TAKE ONE

CARAS: Would you care to speculate a little bit about where mankind as an animal is going and what he might look like in another four or five million years?

MITCHELL: How man will look in four to five million years, physically? Natural selection is acting upon man today in our present environment and, depending upon the changes that we make in our environment and the demands that fall on the body, man will change in form. We'll see the loss of hair. We'll see certain types of increase in stature, different changes in form. I think one of the greatest things will probably be in the next thousand years with its complete integration of the different races on the Earth, and as far as form, I don't know whether the humanoid form through excessive use of machines will begin to lose the use of their legs perhaps: somewhat along the very far-out science fiction line this is a possibility. But beyond that I couldn't really speculate.

CARAS: What might our capabilities be, though? What extraordinary capabilities might we have in four million years more of science? Can we alter our own biological form? Could we change our biological form? Could we have a genetic pool that will still be controlled? What might we be like as people?

MITCHELL: Man, given the guidelines of science as we know it today, biological science in four million years could do a great deal with making man a better unit, a better biological unit. And I believe that in that time if we enter a social system which regulates and can select for, and perhaps with the newest molecular biological investigations that are going on, change genes or modify them and in so doing I believe that the possibility is very strong that man will be able to change himself mentally, his mental capacity. Oh, lots of things are available there. We investigate the sub-molecular units.

CARAS: Do you suppose then that if there is extraterrestrial life in the universe that, and it is very much older than ours, some most certainly would have achieved these enormous capacities and capabilities?

MITCHELL: I believe that. Speculating on the existence of extraterrestrial life somewhere far out in the far reaches, and I do believe that this occurs, and if they are older civilisations than we are that one of the necessities for contacting these planets, probably somewhat similar to ours, would be an interchange of ideas and of technology and that they would be tremendously far advanced from us.

CARAS: Do you think that, assuming again that there is intelligence elsewhere in the universe – do you suppose that any of that intelligence has visited us in the past? I'm not talking necessarily now about flying saucers, but seriously, do you think as there is an awful lot of talk about this? What do you feel about it?

MITCHELL: I believe that if there are civilisations in other parts of the universe so to speak, that have mastered flight beyond or approaching the speed of light that we have been visited, that they must have some means of assaying as we are doing now the measurement and the totalling of stars and their numbers…

CARAS: Do you feel this is something we could profitably do ourselves?

MITCHELL: Well, as far as travel, right now, no, at this particular point. But as far as communication, as I said before, in several areas different laboratories have begun proper programmes of analysing radio data and are actually sending messages, perhaps. The problem has been a coded message I think, as far as I can recall reading.

CARAS: How will it be possible for man to travel vast distances in the cosmos? The universe first, of course, then the cosmos?

MITCHELL: Travel beyond our particular planetary sphere at this time will have to be done on a generation basis. In other words, if we were to have to migrate now we have the capabilities of building spaceships we would have to plan on generations and we would have to supply the ships in that form because of the tremendous distances involved and the speeds at which we can travel. Even if we could travel approaching the speed of light.

CARAS: What about hibernation?

MITCHELL: For limited time periods. I believe that hibernation would be a means of combating psychological problems of space travel, limited oxygen, limited water supply, limited food within periods of six months. Perhaps to Venus this area of travel at the speeds that we are approaching now.

CARAS: Can you give us a description of what hibernation might be like? Sort of paint a picture for us? Would they be carried on board unconscious and what would their respiration be like and how will this condition be maintained? How will it be awakened? What might a hibernation procedure of the year 2001, for instance, be like?

MITCHELL: You'd like me to speculate on hibernation for humans?

CARAS: Yes. Rather, extrapolate.

MITCHELL: Well, based upon our present knowledge of hibernation and speculating what may come of future investigations; biological investigations of hibernators and the use that hypothermy has been put to at this time in modern human medicine I think it is well to assume that we could, in the future, place man in a state of hypothermy approaching hibernation, and so provide him with mechanisms within his spaceship. This is where we're going to travel for awakening at a pre-set period maintenance in travel; the maintenance would be almost nothing – once placed in a spaceship these people could travel for two or three months and be awakened by the integral computer in the ship, by simple ejection or incubation warming.

CARAS: How is a man initially put into a state of hibernation? Is he given an injection that causes a biochemical change, such as occurs in a ground squirrel in the fall?

MITCHELL: Well, I think the change here would be one we would have to go into the endocrine system and probably into the hypothalamus and we would have to, undoubtedly by endocrine manipulation, produce the general overall metabolism as a natural hibernator is capable of doing. Mechanisms we don't understand.

CARAS: What would the body temperature, the respiration and the pulse rate probably be?

MITCHELL: Well, the temperature would be the temperature of the environment: man could probably be maintained if he could be depressed in an endocrine fashion, electronically depressed as a natural hibernator. This could probably maintain around 10 degrees centigrade with perhaps even fibrillation of the heart of a limited heartbeat, just simple muscular fibrillation. Respiratory rate would probably be imperceptible.

CARAS: Can you think of any change that could possibly be made to the human body by surgical, if you will, manipulation to better suit him for space exploration and is this something that might ultimately be achieved once it was deemed advisable and good by other means? Would this be initiated by an injection of some kind?

MITCHELL: Would this be initiated, would hibernation be initiated by an injection?

CARAS: Yes. I want to hibernate, now what do you do to me? Push me into a refrigerator and slam the door like Hansel and Gretel?

MITCHELL: No, you'd have to prepare him, I think, endocrinally beforehand. A man would have to be prepared. The general activity that we see in a normal hibernator is one of reduced metabolism rate and it's a general overall activity brought about undoubtedly by the animal with some kind of an internal clock mechanism and in this way when fall comes the animal goes into the burrow and begins to sleep. Perhaps longer each time period and the sleep deepens, enters a lethargic condition and finally into deep hibernation. Now even in natural hibernation this is intermittent, with periods of awakening at intervals. Probably determinable by what the environment is like. Now, this could be brought about in man by, I believe, in the future by injections, perhaps by suppression of the pituitary or of the hypothalamus or maybe it might involve implantations in the hypothalamus to depress the temperature and general metabolism.

CARAS: There has been a lot of talk about the development in the foreseeable future of the super-intelligent machine, a great-grandchild of our primitive computers. Some people have suggested that this won't be an electronic development, this will be a biological one. That super-intelligence, machine intelligence, will be achieved in a biological machine. Have you given any thought to this?

MITCHELL: No, I haven't.

CARAS: What about biological machines in general? Is this an area of any particular interest to you?

MITCHELL: No.

2001 opened at the London Casino Cinema on Old Compton Street

Philip Morrison

[Department of Physics, Massachusetts Institute of Technology]

4 MAY 1966 / SCENE ONE / TAKE ONE

CARAS: Mr Morrison, what are the chief arguments that militate against there being life on planets other than on Earth?

MORRISON: The only argument against life on other planets is just our uncertainty, our ignorance. Every analogy and every argument of inference leads to the conclusion that there must be such life but since we have no real theories, no confirmed theories of the origins of life or detailed nature of its evolution, naturally we cannot be sure. But it seems to me the preponderance of the inferential evidence is all for it.

CARAS: What are the chief arguments in support of there being extraterrestrial life?

MORRISON: It seems to me that the most impressive one is given by a photograph – well, it's not an argument, it's a sense of conviction – is given by looking at a photograph of a distant galaxy, a reasonable galaxy like the Andromeda nebula. You see this great big lens of stars. Perfectly bland photograph with its whirlpool structure. Now we know that we live in exactly such a galaxy, one spot in that is where we live. There is no arrow in the sky where we live. It's just a commonplace part of that structure. Here is another structure just like that with a thousand million or maybe some hundreds of millions anyhow, of stars very like the Sun – distinguished in no way from the Sun. It seems to me absurd to think we are unique when the essential circumstances of our existence are multiplied so many times.

CARAS: Can you envision life arising anywhere and not following an evolutionary process? Could life be static on any planet?

MORRISON: No, it could not. Life could not be static. It has to evolve according to, as far as we see it, in an effort to maintain itself under the changing circumstances and, in general, an effort to extend itself to take advantage of more and more ways of making a living.

CARAS: Is intelligence the inevitable result of an evolutionary process?

MORRISON: I think we don't know enough about the evolutionary process to say whether or not it has any inevitable results. We have the evidence before us that intelligence of manipulating the environment is one way, a successful way, of making a living and it appears to me that wherever life can find a way to do that it will eventually manage to do so, but to say we know it is, I think, a mistake. It might be inevitable, but we do not know that.

CARAS: The structure of the universe as it is understood by us would seem to indicate that there could be solar systems and individual planets very much older than our own. Might we then speculate and think there could be intelligence elsewhere? If there is intelligence elsewhere would it be very, very much older than ours and therefore very much more sophisticated?

MORRISON: If you ask the question what is the age of any existing evolutionary pattern in the galaxy I think you would have to come to the conclusion that its age must be generally as great as ours, perhaps somewhat greater than our own. The issue about the evolution of intelligence is, I think, simply this one: the evolution of intelligence is a very rapid matter and the evolution of environmental control which depends upon it is still more rapid. The history of life on Earth is at least three or four billion years old, yet it is only within millions of years at the outside that primate forms of the sort we recognise as able to manipulate the environment came into being extensively. This is one part in a thousand or one part in two thousand. Therefore, even if the ages of the evolutionary patterns on different planets were only similar to our own – I think they will be – I don't think they would be very much older.

That is not the way to look at it. The point is that the scale of evolution is so long compared to that special part of the scale which is the evolution of environmental control, but unless they started with absolute synchronism, which is absurd, there is no reason for them to synchronise. To start from the word 'go' all over, the natural drift will provide very great gaps in time, a gap very large compared to one part in a thousand. Say, one part in ten, which is still not a very large fraction, but that one part in ten would be a thousand times the entire life span of our species and many thousands of times the entire life span of other civilisations. The point is that the growth of intelligence and environmental control are enormously rapid things on a cosmic scale. After all, agriculture is only ten thousand years old. Suppose that had started five hundred years ago? Well, there is no question of it. It would have taken about ten or perhaps twenty or even thirty thousand years to evolve from agriculture say, to industry, but then there are the other 200,000 years in the presence of industrial technique. That's only a difference of one per cent in the whole slide of the time pattern, so I think it would make a very strong argument that if there are a dozen patterns of intelligent life, half a dozen would be way ahead of us and half a dozen would either not have evolved at all – if that is not too much of a contradiction – or be well behind.

CARAS: What might some of the capabilities be of an intelligence that had half a million years more than we have had to work?

MORRISON: The capabilities of a really evolved intelligent community, of course, are very hard for us to foresee. I am somewhat conservative about this matter. Some people think, I think most particularly some of the Russians and Mr Dyson, that such beings could control in practise the energy of a star, even a whole galaxy. I do not think this is likely. The kind of organism we are talking about, small organisms and not something gigantic on the scale of a star, which I don't think we are talking about at all. Talking about organisms of the size we are, anywhere between dozens of centimetres and a few metres, that scale of life will never be able to dispose of anything much more than a few planets' energy. Now, that's a lot of energy, it's enough to make a great many changes in the surround but I don't think it will enable them to change stars very much, or to change galaxies appreciably – not at all to make a park out of a galaxy. To make a park out of a planet? Yes, quite easily. To double-deck the Earth? Easily. To occupy the oceans? To travel from space to communicate? Sure, no trouble at all. We do that ourselves. But to move planets around or to enclose the Sun with a shell or modulate the Sun and make it send out strong signals, I am rather sceptical about that. Could be, but I take conservative views. I wouldn't like to extrapolate that far. Maybe we could add very brief, bright laser flashes.

CARAS: What are your views about the origin of life on Earth or elsewhere? Do you subscribe at all to the panspermia theory? What are your views?

MORRISON: No. I suspect, and I think the reasons are rather clear, that life is the most interesting and subtle of all physical and chemical phenomena and is likely to arise in the course of time whenever a system appropriate for it is held for some time with a source of free energy. Maybe there is some special likelihood that there is spread of life from one centre to another – except possibly deliberately. I think that is a possibility. But then I think we would not see the long evolutionary history that we have seen in our planet.

CARAS: There has been a lot of talk, a lot of hysteria and a lot of nonsense about visitations to us – not necessarily talking about UFOs but [?] in Russia has published photographs taken in caves[1] of creatures drawn forty-five

---

1] Caras is here referring to the frescos from Tassili-n-ajjer as mentioned in Footnote 2 in the Frank Drake interview, *supra.*

thousand years ago with space helmets on, or appear to be[1]. Do you believe it is possible that Earth has ever been visited by a superior intelligence?

MORRISON: I would have to say that it was logically possible that Earth has been visited by superior intelligence, but I think it is most unlikely. I put myself in the place of that visitor. I would leave unmistakable signs and until we find those signs I think we will have to conclude that there has been no visit. Should those signs appear, I'm willing to change my mind. But I wouldn't let it depend on legends, or the scratches of cavemen. I wouldn't put up with hints… Let me leave a beacon or a monument, a true marker, a note in a bottle, and that is what I think one has to find.

SCENE TWO / TAKE ONE

MORRISON: The matter that impresses me is to look at one of these marvellous photographs that they take with the big telescopes of the nearby galaxies – say, the Andromeda galaxy, our nearest neighbour. There you see this big whirlpool of stars. Now, we know that our galaxy is a very similar structure to that one. That one you look at, it's perfectly in harness. There is nothing there except a bland whirlpool of stars. It's the same thing here. There is no arrow in the sky to show where the Sun is and yet here we are living on this planet. It's very hard to believe that in those thousand million stars of a similar sort, near one of them, in that other galaxy say, which looks just like our own, there is not somewhere, probably in many places, a similar solar system, similar planets, some kind of form of life.

CARAS: And the other thing we were talking about was the origin of life.

MORRISON: Of course, we have no adequate theory of the origin of life – nothing quantative – nothing verified. But everything that we seem to know gives me the strong impression that life is the most remarkable and the most subtle of all the physical-chemical processes and it seems to me that any system with an adequate supply of free energy, the right circumstances in terms of temperature, water and so on, is going sooner or later to produce those initial stages which we will recognise as the incipient forms of life. Then the processes of evolution will go ahead, depending on the environmental circumstances, limited here or there by whatever happens in the surround to develop to an end which we don't know.

CARAS: You've said that you were doubtful about visitations and all the claims thereto because of the lack of artefacts. What about radio contact? What about projects like Ozma[2] and others that have been proposed for a concerted listening programme?

MORRISON: Well, with my friend Cocconi we were the first people to publish, in the last few years, a proposal for such a listening programme. I fully believe that such a programme is a sensible thing to do. I do not think it is something that needs to be done urgently, something that needs to be done hastily. I think that after mature consideration of what are the best circumstances, the best frequencies, the best media, the best channels, the best directions, after there is a scientific consensus of all the probabilities, then I think we should mount an international, a rather long-range, slow, careful search of the stars, the nearest million stars or so to see if anybody is not sending to us to pick up some sign of a galactic community of intelligence which very possibly exists.

CARAS: We would pick up radio waves even if they were not intentionally trying to contact us? If they were using radio at all, would we not?

MORRISON: It is very difficult to pick up radio waves at all if they are not beamed at us. Of course, since we don't know what the technology amounts to we cannot be sure of that statement. But my guess is that every such community of intelligence is interested in picking up new members. That is the most entertaining thing they can have, a primitive new member with his raw, fresh ideas, his plays, his history, something that cannot at all be predicted by them. They know all about stars and planets and probably even the general features of evolution. What they certainly do not know, because the class is simply too large, is the works of the mind, it is impossible for them to know. To predict in any kind of detail, that's fascinating for them. They must want novelties of that sort. So, I believe, there is a genuine impulse on the part of such long-lasting societies, if they exist at all, their impulse would be to make other contacts. Probably they get PhDs in anthropology for picking up a new planet. So I think they are beaming to us along some channel which we haven't yet quite found.

CARAS: Should we be propagating?

MORRISON: Whether we should send or listen is an engineering question, a question of engineering expediency. I believe that it is not sensible to send,

---

2 Project Ozma was the first organised attempt to listen in to the heavens for artificial radio emissions, 1959-60. It was started by Frank Drake who was then at the Green Bank Observatory in West Virginia. It was the mother of all later attempts. The name comes from L. Frank Baum's *The Marvellous Land of Oz* (1904) wherein is Princess Ozma of Oz.

which is a more difficult task than listening for a time… I think maybe in decades or centuries, we should start sending on the grounds of 'perhaps everybody is waiting for the first word?' That is a conceivable situation, but much more likely, we are simply rather amateur about the whole matter. It is much more difficult to build a transmitter than it is to build a receiver. Everybody can own a receiver in his house but few people own powerful transmitters. From that analogy, we should listen sensibly for a long time, until we are convinced that it is very hard to do better, then we can try sending.

CARAS: Could we be convinced or assured that they would use radio?

MORRISON: No, we cannot. We need a consensus, a scientific consensus on the kind of channel they will use. I personally have given what I think are strong arguments still in favour of the radio channel as the one rational choice. I'm not convinced those arguments are secure. What I want to see, therefore, is years of work with many people thinking about it to see if there are not one or a few channels which suggest themselves as the optimal channels to use. When we find those, pick on them. So far, I know of only three possibilities that anyone has suggested – perhaps four – the fourth being rather unlikely: radio in the thousand megacycle region; optical light using coherent laser light, high-energy gamma rays, and neutrinos of all of these. I still think much the most practical is the thousand megacycle radio. You recognise that that's where nature makes the least interference and this is the basis for our saying that it's rational. The noise is least in the radio channel.

CARAS: Science fiction writers for a long time, and a lot of other people as well, have made quite a fuss about whether there would be a danger to us from extraterrestrial sources. Someone has even suggested that if we did receive a message should we answer them? This seems to me to be an extension of our own neuroses and insecurity. Is there any reason why we'd find any sophisticated society with a million years on top of our own being inimical?

MORRISON: No. I think that there is no reason at all for the position that says that there is great fear, great anxiety about answering some signal we might hear. I admit it is conceivable that these signallers might be hostile persons, hostile beings, who would eat us, say. My only argument is the following: the task of doing physical damage is so much heavier over the distances of space, so much heavier than the task of making yourself known by signals that if they are capable of doing physical damage they already know about us. They don't have to be told by our reply. They have got patrol cameras out, to patrol vehicles that locate all their food, so they don't have to worry. They'll just eat us up anyhow. So, the mere answering is nothing. The fact that they are sending a signal would be the best sign to me that they really want to make intellectual contact, not something voracious or hostile. If those fellows can do it, they're so far ahead, they don't ask for you to talk. It is as though we had to worry about ants signalling. If you saw an ant signal at you, would you be more likely to step on him or to step on the other ants who didn't show any signs of such intelligence?

CARAS: What do you think the reaction would be – the cultural shock level would be – if the [?] traveller came out tomorrow and said, 'Contact definitely received'? Trust he was right, what would the reaction be?

MORRISON: If the time comes, and I suspect some day it will come, and we will learn about some external community, I suppose this would be one of the greatest sources of new thought in the deepest framework of human thought that we would have had. Comparable, I should say, but greater than the discovery of ancient times in the 19th century, when men began to realise that human history was much longer than they had thought. Perhaps it is even comparable to the intercourse between medieval Europe and China, which restored to them the vision of a new and great world to the East that they had lost sight of since the fall of Rome? Something of that sort, probably still more pervasive in many ways which would be impossible for me to detail, but affecting, I think, all kinds of thought from the most austerely moral and philosophical religions down to the severely practical planners.

SCENE THREE / TAKE ONE

CARAS: Can you envision anything happening that would create greater excitement in the scientific community than the final contacting or discovery of extraterrestrial life?

MORRISON: It seems to me that contact with extraterrestrial life, if possible, would be the biggest news not only in the scientific field but also in every other kind of field for a long, long time to come. I certainly can't imagine a bigger scientific surprise than that. A very much more modest thing which I expect to happen will be results coming back from some NASA surveyor or some equivalent Russian shot in five years or seven years, with authenticated evidence on Martian life. Not jazzy or exciting from the point of view that it will be intelligent, but I think there will be an interesting ecology there, which will be a new biology, like a whole new biological science. I think that is possible in five or ten years.

CARAS: What do you think life on Mars might be like?

MORRISON: Life on Mars will be something like lichens or mosses: rapidly growing little micro-organisms that spread quickly over the planet when the season is good, darkening the planet and disappearing again. Some rough little structures a millimetre or two, little balls, little leaves, something like that, covering the planet in sparse but widespread plains like the plains of the Mojave Desert when the spring comes. Something like that.

CARAS: Sir Bernard Lovell has been very vigorous in his complaints about the Russian programmes. He does not feel that they are taking sufficient care in sterilising. For instance, what are the problems associated with cross-contamination?

MORRISON: I am no biologist, I am much less excited about the problems of contamination than the biologists are. I don't feel that the kind of things we are going to see are so similar to our own that the presence of a contaminant will destroy the data completely: not at all. It may make a little problem but I don't think it's very serious. I would rather not take a heavy risk in losing the mission, yet take a larger risk in sterilisation. I know the biologists don't agree with me, and that will get me excommunicated, but I think they are, if anything, overly careful. (I don't know what the Russian procedure is. I understand the Russians are quite careful too.)

Last week or so I heard of a rather detailed Russian account of the sterilisation of their vehicles towards Mars and I thought it was quite adequate. So, if anything, I feel that all the people in this business have been more than careful. It's true, they tell me, about the Australian rabbits but I don't think – it hasn't ruined Australia. It's a modest but genuine problem and I don't think that the rabbits found Australia much more convenient than the Florida or Crimean lichens are going to find the Martian climate.

CARAS: What is the future of man's space flight? What comes next? What do you see, looking into your crystal ball? What do you see the next stage is? What do you think we will do? Is there any kind of timetable in your own mind?

MORRISON: Well, I'm kind of caustic about man's space flight. I think it's a mistake so I don't think that manned space flight can do anything that unmanned space flight cannot do. It is mainly a symbolic thing, like exploring the South Pole and people always wanted to do that. I'm not against that, mind you, but I haven't myself thought through what the steps are because I'm not close to that kind of physical exploration – adventure – and so on. Naturally, I would like to see it. I watch with excitement the flights on television, but I think it's a great spectator sport. I approve of it in that sense but it's not much more than that. We can do just as well with servo links as the astronauts can do, and a lot more safely. I think perhaps the first time when a justification for sending people out into space will appear once we are doing complicated investigations so far away that the linkage time to bring the signal back to Earth is an appreciable problem for the motion of the vehicles. That is perhaps as close as Jupiter, but not closer. The Moon? I don't think there is any great problem, anything we have to react to within a couple of seconds on the Moon is going to wreck the apparatus anyhow. I don't think a man will do much better and the cost of human safety insurance development seems to me out of all proportion. The risk is to the brave individuals. I would rather see a controlled robot. I'm sure that people can build it now. You go down, do your job in Houston, and spend eight hours being on the Moon. That is to say you see and you feel just what you would see and feel if you were on the Moon. You have apparatus on the Moon responding to your hands and to your eyes, giving you the same signal. You work at it eight hours and go home, and someone else comes and takes eight hours. Nobody is ever in trouble. If the circuit goes bad, if you cannot fix it with the repair apparatus which you can control and have there, then you cannot fix it. That's the way it is, yet no one is dead. It's cheaper because you don't have to supply the air and the water and all that the poor astronaut needs. So, I really think that astronautics is a kind of daring, symbolic thing and I am very proud. I admire the people who do it but, from a scientific point of view, it has no great claim on my interests.

CARAS: The manned or unmanned investigation of deep space. Do you think we will ever be able to, with the problems of time and distance within the lifetime of any one individual? Do you think we will ever exceed the speed of light? What are the problems here in this general area, deep space?

MORRISON: I believe that there is a complex of arguments – none of them certain, all of them, I admit, open to some question but their tendency is to lead in one direction. Namely, that flight over distances large compared to interstellar distances is impossible or next to impossible. That is to say, so prohibitively difficult compared to what you can gain by other channels, so that it will never be undertaken, except again possibly for some symbolic reason. The way in which statesmen fly: the head of the one state goes to the head of another, and once he doesn't spend much time doing that, he travels, sees the country, shakes hands, goes back home again, to represent some kind of rapprochement, some détente. I could well imagine that being set in motion over a period between two communities that have been in contact for hundreds of years. They finally exchange one party, just to make that gesture of community. That is a casual thing, but even as a serious exploration venture I don't think it is wholly sensible. I think that machines will do better than man. Man should stay home and work and enjoy himself, but not go out into space, I believe, except the occasional – the odd brave fellow who will do this.

CARAS: What I meant, I didn't make myself clear, I meant what about unmanned? I said either manned or unmanned?

MORRISON: Oh, I'm sorry…

CARAS: What about sending a machine out and a man going to his desk in Houston if the thing is going to take four million years to get here? Houston might not be there?

MORRISON: Houston will certainly not be there. But this cannot be helped. You see, I think that interstellar communication also is most likely to be only one way, in terms of the lifetime of any one individual. That doesn't bother me. Consider Greek society. It's hard to imagine a single body of knowledge more influential for the formation of the ideas in the Western world than Greek society. There are one or two others. The religious traditions of the Testaments, but apart from that it certainly is Athenian society. Now, no European of Renaissance times up to our time ever had any hope of having a two-way communication with Socrates, or Pericles or Thales or anybody else. That did not prevent us from gaining enormously from that body of manuscript that we have. That is the way I think we will regard the communications that come from space. I don't think that we are going to find them nearby that we can get answers to questions, except with great effort, preparing to wait over the centuries. I do not believe in travel faster than light.

CARAS: Well, then, we will never get an instrument – we can only get a testament from an inhabitant of Planet X if there is someone on Planet X, but we'll never get them to talk or look at Planet X, unless we send something there?

MORRISON: Unless he sends the message of what Planet X looks like, which I think he is doing. Now, if Planet X doesn't have people on it and you want to find out how it looks, which I think is not quite so interesting. You have already seen Planet Y, one that is much closer and has such similar rocks and air. Then you have to send a device, a probe, there. You'll simply have to wait until it gets there.

CARAS: You don't envision ever travelling faster than light?

MORRISON: I do not. Again, every statement I make is subject to some error. That's what we have learned in science, but if we've learned anything, we learned that the nature of space and time that we now seek is broad enough to take in a variety of experience to which we see no exception. I don't really feel that this is likely to be broken. It might be. I would be happy to be proved wrong, but for now I'm unlikely to believe it.

ON DISTANT EARTHS: Philip Morrison, 2003

In nearly every response to Roger Caras in this text of 1966 I imply my need for – and hope in – some evidence strong enough to build upon. We have it now, but as so often in science it is a foundation, though not yet an answer.

On 10 October 1995 two Geneva astronomers announced their fully credible finding of a planet not related to our own Sun. By now we have found a hundred of them, none further than about 100 light-years. They orbit stars very much like our own Sun, but though we know the Sun-like stars pretty fully, we cannot yet detect the exo-planets, unless they are Jupiter gas giants, some reaching 10 or 15 times more mass than our Jupiter! We cannot even say whether or not any Earth-like rocky planets orbit those distant suns, as Earth joins Jupiter in our home system.

No earths to report, but none to exclude!

As always, we await an answer with improved efforts. By 2006 or even earlier we should detect Earth-like exoplanets – if they are present.

What we have is a lot of systems but none that could seem home-like. But we cannot exclude Earths in these systems either! We don't know. We wait for an answer when our better detection arrives. That will be a great day – maybe. It's hard to wait.

SK and Arthur C. Clarke

A. I. Oparin

[A. N. Bach Institute of Biochemistry, Academy of Sciences of the USSR, Moscow]

10 JUNE 1966 (MOSCOW)

[TRANSLATED FROM THE RUSSIAN]

The origination of life is not an extraordinary event, a lucky circumstance, as has been a general concept until quite recently. It is an inevitable phenomenon, part and parcel of universal evolution. In particular, our terrestrial form of life is a result of the evolution of carbonic compounds and multi-molecular systems formed in the process of this evolution.

The initial stage of this evolution, the emergence of primitive carbonic compounds, occurs very frequently in the cosmos. We observe them everywhere, on all celestial objects within our reach, on the surface of stars, in interstellar gas-dust matter, on the surface of planets, meteorites, comets.

On the Earth this stage was connected with the very formation of our planet.

Further on, these most primitive carbonic compounds on the Earth surface grew more complicated. In the primary atmosphere, which had been of regenerative nature and had not contained oxygen, there formed more complicated compounds, highly molecular compounds such as nucleonic acids or proteins. Thus, in the waters of the primeval ocean there emerged a solution of complex organic substances, which are now part of living organisms.

In this so-called 'primary broth' multi-molecular systems made their appearance. In this respect, mention should be made of the research done by Mr Fox, on the so-called micro spheres. In our laboratory we found out that as soon as the complication of molecules had reached a certain degree, multi-molecular formations emerged called conservate drops. These drops obtained in our laboratory may serve as model of pre-biological systems whose development and complication led to the emergence of initial living organisms. In particular we could prove the possibility of the origination in these conservate drops of metabolism, of pre-biological natural selection of the conservate drops and, thereby, we could show the path along which the process of com-

plication of these multi-molecular systems took place, as well as the process of their transformation into primitive
living organisms.

Further, Darwin's evolution took place, the evolution from the primitive living organisms to more complicated organisms, which has terminated in the emergence of the human being and human society. Could similar phenomena occur on other celestial bodies? As I have already mentioned, we can observe the initial stages of evolution everywhere, on various celestial objects. The second stage is also quite likely to have taken place elsewhere in the cosmos. There can be no doubt in the existence of highly molecular complex organic substances on such objects as the Moon or Mars. Even if these substances could not be formed on those planets, they could be brought there by falling meteorites.

For the third stage of evolution to have taken place here on the Earth large expanses of water were necessary, which we could not observe and did not even expect to find on the Moon or Mars.

However, in the process of evolution of these planets, at the initial stage, these planets, for instance Mars, could be more rich in water and life could have emerged there along the same lines as on the Earth and, having once emerged, it could develop and adjust to those severe, I would say unbearable for human beings, conditions existing on these celestial objects.

The knowledge thus obtained will be of great importance to our perception of life. No matter what we find there – organic substances, remnants of some living organisms that had existed there and vanished as a result of conditional changes, or a highly developed form of life – in any of these instances the knowledge of this type, the date obtained, the very encounter might be of great value to us in our perception of terrestrial life. We possess, so to say, a single copy of the Book of Life, that of terrestrial life, while the knowledge of other forms of life could tell us about our past and, what's more, it could supply us with many clues as to our future. The discovery of new forms of life superior to ours would immensely enrich our culture and expedite our development.

Thus, human venture into space, direct perception of the solar system and, in particular, of Earth-type planets will add much to our perception of life and its development.

I doubt if we can seriously talk of any visits to the Earth from outer space that could have taken place in the past. It is still the sphere of science fiction more than that of science. Of course, science fiction is fine in its own right. However, we should in all honesty say that the boundaries of our knowledge are too far from the point where we could seriously discuss this problem without having any evidence whatsoever of communications of this nature.

Evolution of matter is not a single straight line. It is, rather, a bundle of divergent rays, a fan, and the origination and development of life is one of the rays of this evolution, one of the roads on which the transformation of matter from primitive to complex forms has taken place.

Doubtless, a great number of complex and highly developed forms of the evolution of matter are to be found in the limitless expanse of the universe. But it is in no way imperative that we call these forms 'life' or consider them such, as they differ in principle from our terrestrial form of life. It is my opinion that should we come across such phenomena in the process of our increasing space effort, we can work out some name other than 'life' for them.

It should also be borne in mind that life emerged as a result of the evolution of carbonic compounds…could develop along similar lines in the initial stages only, even in the case of closely resembling planets, say, of terrestrial type. Peculiarities of this evolution, for instance, at the stage of Darwin's evolution, are closely related to the conditions under which the evolution of the whole celestial body in question takes place. And we can hardly expect to find anywhere in space human beings or living organisms morphologically similar to those of our terrestrial world. I presume the highly organised forms that may be found elsewhere in the universe are completely different in their appearance, which does not, however, rule out the possibility of finding intelligent life of a new type other than our terrestrial life.

Taking into consideration a great number of planetary systems within our galaxy alone, there exists a strong probability of finding one or several planets similar to ours, being an almost exact replica of our Earth. However, the development of life is such a complex process that, even in this case, full coincidence of the forms of life on these duplicated planets with our terrestrial form of life is hardly possible. These forms may be very close; however, certain distinctions will be observed.

SK with his daughter, Anya, on the set

Harlow Shapley

[Harvard University]

4 MAY 1966 / SCENE ONE / TAKE ONE

CARAS: Doctor, what arguments militate against there being life on planets other than Earth?

SHAPLEY: In our own solar system the thought is rather clear that we aren't going to have life in this solar system because of the temperature problem. We can't have life without it being metabolic or metabolism working and it's not going to work if the water is all frozen or is all steamed away. We need water in a liquid state and we don't think we have much of it in this solar system and so that is the argument that militates against having life as we know life, as we practise life, in this solar system. But there are a lot of other systems which we could have life on. In fact it would do for me to point out that our studies of the number of galaxies and the number of stars in galaxies in all lead us to the conclusion that there must be something like ten to the twentieth – at least ten to the twentieth power stars. Ten to the twentieth, that's a pretty big number, but if you write down one and put twenty zeros after it that would be the way it would look if you were expressing it in numbers. Well, that's a whole lot of stars and suppose we say that only one in a billion, one in a thousand million, is suitable for life, we'd still have a tremendous number. We would have a 100,000 million galaxies and stars that might harbour life. The chances of life, of course, existing in here is very high because just the fact that we exist on this planet and get along comfortably on this planet in itself suggests it would have duplication and multiplication elsewhere. It would be very odd wouldn't it if we had a 100,000 million billions – gosh! – I said all of those numbers – but that's what it is – ten to the twentieth power stars is one with twenty zeros after it – and that's 100,000 million billion. Well, with all of those suns a great many of them, perhaps a third of them, would have planetary systems around them. Now, most of those planetary systems might not be very comfortable at all, but if only one in a million were comfortable we would still have an enormous amount of life. So, I think just from the statistical standpoint of that great abundance we have of stars, this ten to the twentieth I mentioned; that's a minimum by the way, and the abundance we have of these galaxies and what we know about the way stars might form and planetary systems form we can't escape the idea that there must just be a tremendous lot of life. It would be very vain on our part to think that this particular planet which goes around a run-of-the-mill star, to think that that would occur only in this system is ridiculous. We are peripheral – I like to use that word. I mean we are on the edge. We are not in the centre of a galaxy, we are not important in a galaxy, we're just here.

CARAS: Could some of this life be very much older than our own?

SHAPLEY: Well, that's an idea, isn't it? It very likely would or could be. There is no reason to think that we are [?] in the time sequence so I think that it's very likely that it could be life that is older and life that is maybe just getting started in many other systems: Yes, so I think the idea of life older than ours is a pretty good thought and I would say the answer is very probably yes, there is a lot of life.

CARAS: Would you speculate or extrapolate what might life be like if it has four million and ten thousand years of civilisation to become sophisticated, rather than the ten thousand years that we have?

SHAPLEY: Ten thousand years of life, of course, ten million, maybe ten billion, yes. That will be available and would it become some of it so sophisticated that it's quite unrecognisable as life? I think not. We have evidence from our statistical examination, from our study of the biochemistry. We have an indication of life being an automatic, almost a natural consequence. It comes the same way as there's an evolution of stars or an evolution of galaxies. The evolution of the chemistry on the surface of this planet wouldn't necessarily be such that we would say that that's life. Of course, we ought to define life, shouldn't we? We should define life – and many people take shots at it, but I think the simplest one is to say that life is an activity of replication of macro-molecules. That sounds a little hard – I'll say it over again. The macro-molecules when they divide up are self-replicating and represent what we would call life's metabolic operations. And we should always remember this – there are a million kinds of life on the surface of the Earth – a million kinds. That suggests that there would be lots of things that could occur in there that we've never seen, never would see, but with a million kinds of life we could boast a good deal and be quite sophisticated about whether any particular metabolic operation that we call life exists anywhere. Even on Mars there are goings-on no doubt and in general I think the cautious astronomer would say that there may be algae. There may be fungus, fungoidal growths there of some kind, but nothing very tremendous like here or even like the dinosaurs. They could be very simple sort of forms, and why simple? Because if they aren't simple we are not going to get very far. It's going to be too cold in a lot of places for life.

CARAS: What are your views on the origin of life on this planet or any other planet for that matter? Do you take any stock of the panspermia theory?

SHAPLEY: We have, of course, at times thought that we could have life on this planet without actually having it created here. In other words, that life could be imported from places like Venus and Mars and Jupiter and Pluto. We might bring life in that way and therefore not require some operation on this particular planet. But that has been dropped. We call it the panspermia theory and it was devised and developed largely by Arrhenius, a Swedish scientist, over a hundred years ago. It was largely designed by him because he thought that would be the easiest way, but it doesn't help much because to be sure we can find a lot of opportunities for life, but are we going to have any life unless conditions are some way right? You know, it's only by a very narrow squeak that we are here on this planet anyway. That we succeeded, that our particular forerunners did go in this direction and a great many of them didn't. As one knows from much reading and much writing on the subject, a million species of animal, at least a million, have been on the surface of the Earth and are now gone. In other words, there's a tremendous loss of life. Or, to put it another way, we see that the surface of the Earth by way of the fossils is a graveyard of biological failures that didn't come through. And our ancestors apparently are only of that because we have here Homo sapiens but we don't have much else to show for a million years of evolution, so I think it is by a narrow squeak that we are here in the form that we are here in, a narrow squeak. But then we did get here and by a lot of episodes we did succeed in getting sort of an important place in this whole scheme. On Mars the conditions are so poor that I would not want to gamble that we are going to find anything but we are going to send apparatus out towards the vicinity of Mars. We are going to make a lot of photographs. We are going to do a bit of thinking on the subject and we may come to the conclusion that Mars probably does have a low form of operation that we call metabolic life. Could be.

SCENE TWO / TAKE ONE

CARAS: Doctor, what is the difference in attitudes now towards questions about extraterrestrial life compared to, say, twenty-five years ago?

SHAPLEY: Oh, we are brave now! We were timid twenty-five years ago. If one of my graduate students twenty-five years ago was to say I want to write or study on the origin of life I would ask him to close the door as he went out and do it quickly because you didn't do that. It wasn't proper, especially for the young, to commence speculating in a big way about the origin of life or even the size of the universe, but then the size of the universe, we accepted that after about ten years and it is granted we now know that the centre of our Milky Way system is twenty-five or twenty-six or twenty-seven thousand light years away. Way off in Sagittarius, not near us. We now know and accept that we are peripheral. We are just out on the edge of things going around in 200 million years a central star and it is considerable change we have had in accepting some of these great discoveries. In fact, I am on the verge of writing out and maybe spelling out a story of the ten major achievements of the past fifty years that have changed man's way of thinking and doing things in and about the world we live in. Ten Revelations I call them. You know about the Ten Commandments. Well, this is different. Ten Revelations is what we unravel as we study modern things, and what are the ten? Oh, they are things – I'll just name a few – like knowledge of the origin of life, another one is about cosmic evolution, another one is about some of the horrible things like relativity and cybernetics and medical triumphs and all of those things come in my ten and advances are big. But if a graduate student had tried to do that and put that to me about twenty years ago I would say no, no! Please go back and measure the variable stars and get some information or something. So, you see and I see that we are moderately brave in doing brave things at the present time. There is a little point in there that we are now getting brave enough to discuss and that is cosmic evolution. My idea about cosmic evolution is that everything that we can name – material or immaterial – evolves and changes with time. It goes from simple to more complex or moves away. What do I mean by that? I mean that we have accepted the theory that we know how stars are evolved. The evolution of stars is something we can talk and we write about and some excellent things have been done from the Harvard Observatory on that particular subject of the evolution of stars. We also know about the evolution of nebulosities of the hazy stuff in the sky. It has been known about since the days of Immanuel Kant but we did not know just what they were. Now we grant that that one stage in the evolution of stars and stellar bodies is these nebulosities we see, like the Orion nebula, but we also can arrange the kinds of galaxies in a series and so we would say that definitely the galaxies also were tied up in this evolution. And there we are. Now, we know about the biological evolution, animal and plant, that has been accepted and is easy and even a beginner can write and think on a subject of that kind without losing any prestige, but what about matter itself? And there is where the big development has come and in recent years that is, are these matter, these material things? Atoms? Do they evolve? Is there an evolution from hydrogen into helium into oxygen into lead into gold? Is there some evolution going on there? The answer is we didn't think so a few years ago, didn't think so very confidently because there were arguments against it. But now we say no, definitely. We know how to evolve hydrogen into helium. We know how the universe does it. Here, if you actually check up on it you will find that this universe we talk about, the stars and galaxies and all, are most of the universe. They are more than ninety-nine per cent of our universe we work with. In stars mostly are hydrogen and helium, those two elements, and there is one tremendous thing back in there that one should realise and that is that we know the universe is mostly hydrogen and helium at the present time. We know that the hydrogen is being burned up into a helium ash and therefore there is an evolution from hydrogen into helium, but in just the recent years we have come to the conclusion that we know how to go from helium up to the heavier elements up to the bigger elements. To get to oxygen, for instance, and oxygen is awfully important. A human being is two-thirds oxygen. How can you get oxygen, because to go from hydrogen and helium into oxygen takes a lot of energy, a tremendous lot of energy? It takes an evolutionary step of some sort and where would the energy come from? We have the answer – it comes from the novae – stars that blow up in a way that blow their roofs off. They have high temperatures. The temperature goes way up. It goes from ten million degrees up to a 100 million degrees and if the temperature is a hundred million degrees that is enough to boil or to cook, you might say, your helium into oxygen. And so we have found the answer. But that isn't enough heat to go all the way up into iron, is it, and lead and so forth? No, it isn't. But we have the answer there also. What is the answer? Supernova. We had novae you see for that evolutionary step from hydrogen into helium, burning hydrogen fuel into helium ash; and now we realise that in the novae we have temperatures that will take us further up this sequence of atoms of various kinds and in the supernovae we have a temperature in billions of degrees that can do a lot of this cooking. That can go a long way. And so there is something that has happened in the last ten to fifteen years, it's our realisation that we now know how the elements evolve. Well, if the elements evolve that way and biology evolves that way and the stars and the galaxies evolve that way, we might as well go the whole distance and say there is a cosmic evolution. The whole works evolves with time. Time is a factor in the matter. I don't know what the end is going to be. Some people say what will happen when we run out of this fuel? I don't know the answer to that. I don't want to guess at the answer to that. We will say where does the hydrogen come from? You say that hydrogen can be boiled into helium and then to go up that way you can say hydrogen would go in that direction, it is automatic, but where did the hydrogen come from? Hydrogen, the mother of all the atoms, is just the simplest of all atoms, namely, the proton and the electron that make up that particular atom. And there is the element and so we know. Our groping around for some bright ideas: whence came the hydrogen out of which the universe has developed and grown? And we don't know the answers, and if you were to say in answer you, being scientists in general, well, it may come from twists in the space time. Maybe a source we are building on. OK, I say, who is the twister? And then you are stuck again. So, they are big questions that need some big answers in the future.

SCENE THREE / TAKE ONE

CARAS: Doctor, in your book, *Of Stars and Men*,[1] you did a very easily conceived, easy-to-understand, easy-to-follow breakdown of the conditions that life requires on a body, on a planet and you went through this process of elimination if one per cent of them had this and one per cent of them had that I wonder if you could break those conditions down for us now in relation to the planets and the conditions that might exist?

SHAPLEY: Life as we know it, of course, demands various conditions to be able to operate that metabolic machine that I referred to, and one that seems to be most necessary is the presence on a planetary surface of water in a liquid state. If you had water in any icy state where your metabolism is so slow, you don't evolve very rapidly, or you don't grow, you don't exist. If your water is in a steam then again you are in trouble, you can't have the molecular operations and so we have to have those conditions of water in a liquid state available on any planet that is going to support and develop life. I think for instance, that the planet Pluto is so far out in space that it is colder than the very cold and you cannot have liquid water. You would have ice. Your H2O and maybe some of the other molecular things would freeze up. The same is true for Neptune and Uranus and Saturn probably, and Jupiter. It is a little

1 Boston, Mass: Beacon Press, 1958.

less certain about the Jupiter business. But when you get past the Earth, you get to Mercury. That is too hot. So, there is what I would call a liquid water zone or belt that is a necessary planetary situation. But once you have got a planet in the particular region where you can have liquid water, you need to have certain other properties. Your planet, for instance, must go in a more or less circular orbit around its primary star – let us call it the Sun – must be in a more or less of a circular orbit. If it is elliptical like that of a comet and long and elliptical because we have them back where it is too cold, too hot and so that's one condition we need to have. Some eccentricity in the orbit of the body, the planet we are talking about. Another condition we must have is the right kind of chemistry because you can easily imagine chemical situations such that it would not persist. So we have got to have good chemistry and now we know about chemistry a good deal because our spectrum analysis tells us a great deal about the make-up of it. We know, for instance, that the Sun's chemistry is such that it would not harbour life, for two reasons. One is the chemistry is maybe too wrong, that is the actual kind of atoms that are there, but largely because the atoms are broken down. We don't have molecules, and life, of course, is a molecular operation so on our Sun where on any of the other stars or even out in space there is no chance of life. What we call life in any of those. Life is to the liquid water zone, you might say, of the planets that are in the right kind of orbits. Other things that we need to have is to get life started if you are going to have life go on, persist in that way. One of the necessary conditions is you get it started. By golly, we know something of how that is and we didn't when I was considerably younger and all – in the first place we didn't talk about the origin of life. We didn't know enough to talk about it and we felt embarrassed if we would make some remarks that way, but now we know that life is a natural automatic development of matter. You have, we will say, the various elements in a gaseous state on the surface of the Earth. Life is going to appear where the chemical conditions are right. Well, they are right now on the surface of this planet, but back a few thousand million years ago when the Earth was cooling off and rocks would be too hot the life that could get started there would not have some of the necessary elements. What it would have, according to the researches that we have been making in the last few years – it would have methane and ammonia and water vapour and hydrogen gas and maybe one other – all transparent not much – not greater quantities of them. But those were the kind of elements of molecules that were on the surface of the Earth when life began. But now we can have those same elements in a laboratory. So we can set up a laboratory as Stanley Miller did in Chicago. After we pointed out the nature of what was the chemistry for life, Stanley Miller set up the same kind of chemistry as the atmosphere of that time, the same kind of chemistry as there is out in space. That is out in space, of course, you have electrical discharges from lightning and let us say at the primal atmospheres, and you have these that I have just named: ammonia, methane, water vapour and hydrogen gas. You had those out in space but we can have them in the laboratory. That is what Stanley Miller, under the guidance of Harold Urey, and some of the rest of us put our oars in, what we could do also on the surface of the Earth and so we set up a laboratory that had just those things. We didn't have any lightning from the space outside but we could put in electric discharges, which we did, and so here we were. Stanley Miller in one day, he told me, in the first day he could see that his apparatus was showing something pink in what was otherwise a transparent gaseous medium. He let it run for a week. At the end of that week it was decidedly pink and then he stopped his experiment and analysed what he had done. He had used the same machinery that omnipotence, if there be such, could use, and we used it and he tested it and he used the new techniques that we don't know much about – mainly chronographic analysis – and what happened? He analysed it and found he had amino acids and that was one of the big jumps. I would say it was one of the best experiments of the last ten years. With Stanley Miller and since then many repeaters carrying through an analysis of the kind of an atmosphere that was on the surface of the Earth a few thousand million years ago. Because what he found in his analysis was amino acids. First, just two or three and then at the end of the year, all twenty of the commonly accepted amino acids were found in this University of Chicago laboratory experiment. And that, of course, is a tremendous thing to me that I was just so delighted when I heard what happened because there was something, a step toward life because amino acids are what proteins are made of and proteins are what human beings are made of. This was a marvellous experiment and there it was. So, we found out how to do those things and since then we don't need to appeal to the Almighty, I mean, to the supernatural or to miracles for the origin of life. We know the tricks of how it can be done, how it undoubtedly was done on this peripheral planet in this one solar system and probably elsewhere.

CARAS: Sir, can you envision life arising on any planet and not undergoing an evolutionary process? And the second part of that question: if an animal undergoes an evolutionary process, could we assume that at the far end, given enough time, there would be intelligence?

SHAPLEY: You raise the question of intelligence in animals. A million kinds of animals, at one step in the development of life. You use that word intelligence. I'm sorry. I don't like to use the word 'intelligence' because what is intelligent and what is not? It seems to me that this lowest form –say an amoeba, that, or some other protozoa or proto phyla of a billion years ago, that that was intelligent. It would come up with some contact. It would meet a little grain of sand, or meet a [?] or some other low form of animal, and it would choose the one that was digestible. So, that is intelligence. Too often you are trying to make intelligence like human intelligence. But the human intelligence is just one step and we have just lots of intelligences. I think of all biological phenomena, in fact, I am not sure that you would say that when the sodium works with chlorine, that isn't some atoms that are rather intelligent because they make salt.

CARAS: Self-awareness. Is self-awareness at the far end of the evolutionary scale inevitable? In other words, if life arises on Planet X, or Planet Y, will it eventually be aware of itself?

SHAPLEY: Awareness is a question that is raised in there but just what is aware ness? Who is, what is aware? Is a dog aware of himself? Is a fish aware of himself? Is this amoeba of mine aware? It knows that one thing is edible and digestible and another is not. So, intelligence to me is just a matter of degree. You can have high intelligence, meaning complicated business, Freudian evolution and things of that kind and you can have intelligence which is middle-size and low, but I think it is just a graduation all the way through. I don't see any chance of isolating and saying some particular animals are intelligent and others are not. It is a degree.

SCENE FOUR / TAKE ONE

CARAS: Doctor, there has been an awful lot of nonsense as well as a lot of sense spoken in years past and sometimes it is hard to separate them. What about all this business about visitation? I'm not talking now about flying saucers but there have been serious claims, some made by Russian scientists, talking about evidence of extraterrestrial visitations in the past. Is any of this worthy of credence?

SHAPLEY: I think that one should be very cautious in accepting other intelligences coming to here or having arrived here, operating here. I don't know of any sound evidence in favour of that.

CARAS: What about radio contact? Do you think there is a possibility that there has been an attempt by extraterrestrials to contact us by radio and are we doing enough? Should we be doing more to examine this area?

SHAPLEY: We have done so very little in exploring the possibility of getting messages across space. It has been of considerable interest to me because of the people who are excited by this particular operation. Are we doing as much as we should? No, I think not, but I think we are going to do more because, after all, there are several big instruments that are being built up and one of their goals is going to be the check-up. The thing that bothers us, however, is the time element, that comes up there. The nearest planet you could hear from, talking with other people like Fred Whipple and Fred Hoyle, the nearest we come to being useful is trying out stars, the neighbourhood of stars that are not too far away. But the best we have been able to do, so far, is to choose some stars which are about ten to twelve light years away which means that if a message was generated in some way – in another system – and the most satisfactory place that we know would take ten years or more before it would get to us. It is a hard job to get something that will work but now there is a chance. I don't think I have said this before and I believe it pretty devotedly. There is a chance there may be bodies a good deal nearer than the few that have been chosen out as approximately like the Sun and not very far away. That is, those that are ten or twelve or fifteen light years away. They are just as hard to deal with if you could send a message that would be understood, and then expect an answer. It would be, say, twenty years before you would get anywhere. By that time so many things would happen that you would be in it. But there may be, and not many would yet admit it, there may be some closer objects that we don't know about. I will tell you what they are. They would be super-Jupiters. Now the planet Jupiter, of course, is attached to the Sun. But there could be objects as big as Jupiter, as massive as Jupiter, that are out in space and not tied with this solar system at all and it is just barely possible that with our radio telescopes and other ways we may find some other, we may run on to them. So there, I think, is the best hope of the future, in getting contact, is finding nearby objects.

Now down in Swarthmore College van der Kamp and his associates have done about the best that has been done in trying to find nearby planets and they have a number of them that seem to be hopeful. They have about four. But all we need is a flow of time. If we could check it up in a hundred years from now we will find not just four objects that seem to be planets associated with stars, but there may be forty by that time that we know with some confidence.[2] So, there is the pace that takes patience and statistics. Well, what about these nearby ones? I just make the suggestion that if Jupiter were cut loose, or something as large as Jupiter, in space, it would radiate in the very long wavelengths and might be picked up by radio astronomy – might, might not. I have written an article about it, but not taking it too seriously myself I published that article in the American Scholar or in some general magazine and I did not propose it too seriously. But the more I hear of their attempts to get contact with other bodies and I say, let us try for the super-Jupiters, or I call them self-warming planets, because if Jupiter had a good deal more mass, say ten times as much mass, well, then it would be radiating, be rather successfully radiating and then there is another source almost the same. We have dwarf stars. We have sub-dwarf stars. We have stars that are very small compared even with the Sun. What is the chance as we go down we will pick up some that are a thousand times fainter, but still stars? There is a chance. So I think that in the self-warming planets and in the crusted stars, we may have a contact, maybe of something very important, but it is, of course, a dream so far. We will try to do better.

CARAS: What do you think the reaction is going to be on Earth when we finally make contact with an extraterrestrial civilisation?

SHAPLEY: Boredom. I don't know how we would react to the discovery that we were in contact with other objects but I would think that nothing very exciting. There are some suggestions that we may have such bodies, such planetary systems, or such life on a planet, somewhere, running around here all the time now. That, I think, is just dreams. It is nonsense. But then, nonsense is pretty, sometimes.

CARAS: You don't think there would be danger of cultural shock?

SHAPLEY: No. I think not. The Ten Revelations I spoke of, they were enormously important things and I don't think we were shocked by them. We were just delighted with them. We were amazed by them, but the things that come out in man's way of reasoning, in man's ways of measuring and all that kind, they are dramatic. But I don't see shock in it. I don't see much interest in it in fact, except Sunday newspapers.

CARAS: Thank you, Doctor, very, very much.

2 Peter van der Kamp (1901-95), American astronomer, and a leading pioneer in the search for extra-solar planets. He was the director of the Sproul Observatory at Swarthmore College in Pennsylvania.

S. M. Siegel

[Union Carbide Corp Research Institute]

SCENE ONE / TAKE ONE

CARAS: Doctor, what are your views as to the origin of life on Earth?

SIEGEL: The last ten years or so have seen a real revolution in our thinking about the way in which life originated. For many years the doctrine which was put forth in the last century by Louis Pasteur, 'That all life must come from life,' had been held to very firmly. In the 1920s the Russian scientist Oparin put forth a more or less theoretical but very plausible concept as to the way in which organic matter related to life. The building stones of organic matter might have arisen under a special kind of set of chemical conditions here on the Earth and progressed from this rudimentary state into a more and more sophisticated form of organisation, eventually acquiring the attributes that we consider to be life. But from the 1920s on through the 1940s this remained purely a matter of interest, something in a book to be discussed perhaps with students in classes and so on and then passed on. It was then with the work of Harold Urey and his student, Stanley Miller, that these speculations were first turned into experimental fact and it was demonstrated that one could indeed make organic compounds that would serve as the building blocks of living matter out of very simple kinds of substances that would be presumed to be around during the formation of planets.

CARAS: Is there any reason to believe that the conditions that would allow for the beginning of life are exclusive to this one planet?

SIEGEL: The kind of condition that we refer to, it seemed rather revolutionary in the days of Oparin, was one that is extremely remote from the Earth as it is today. For example, the complete lack of oxygen in the environment of forming life elements or life substance was something rather difficult to accept in the earlier part of this century until it was realised that the activities of living matter once the process of photosynthesis that green plants carry out actually had produced the major part of the oxygen in our atmosphere and that the whole progression of life up to the invention of chlorophyll and the abilities that it conferred had been achieved with virtually no oxygen. Now this state of what has now come to be called a 'reducing atmosphere' consisting of gasses such as methane which we know commonly as marsh gas and ammonia and of course the quite ubiquitous hydrogen, is believed to be a primitive state for all the planetary systems within our solar system, and by virtue of the abundance of hydrogen in the universe the reducing environment would seem to be a likely primitive environment for any planet.

CARAS: Do you personally believe that there is life on any planet other than Earth? Most scientists today seem to.

SIEGEL: The question of life on other planets that are or will be accessible to us within the next few decades still has to rest largely on conviction. We cannot prove anything. From a purely objective viewpoint we can infer in the case of Mars, from the surface behaviour of Mars, that there are activities and changes that are correlated with the seasons. It is very tempting to relate these to winter and summer conditions for the surface of the Earth in terms of browning leaves, greening of leaves, as winter and spring activities and so on, but we don't know and we have no facts. So this is why some of us have been led into such laboratory approaches as to try to simulate the conditions of Mars to varying degrees or to put life under conditions at least as harsh as we believe those to exist on Mars and with the argument that if Earthly life can take it, then there is more of a chance that some kind of life is there and perhaps not even existing at the bare edge of survival but even flourishing.

CARAS: Taking now the cosmos. Not our solar system, not even our galaxy, or our universe, is there anything in the world that is known to science that would militate against intelligent life appearing elsewhere than on this planet?

SIEGEL: To go from a consideration of, let's say, insects and lichens and bacteria and bean plants, common organisms of this sort, to a question of highly organised nervous systems and intelligent life, involves a tremendous jump. Whether one is a biologist working with the lower organisms as I do or even an experimental psychologist, it seems to me that there is a tremendous gap here. But in terms of evidence there is a reasonable picture I would say of what gives life an advantage is evolving intelligence. As a pure matter of competition, intelligence can replace strength. Intelligence can replace bulk. Intelligence is a good item of the making by natural selection for survival and therefore it could be assumed that intelligence would have a potentiality for arising under some set of circumstances, just our ignorance of our evolution. What specifically directs animal life towards intelligence we don't know. But it happened here and this makes it at least possible happening in the universe as a whole.

CARAS: If life arises on a planet, is it possible for it to be static or is it bound to undergo an evolutionary sequence of some kind?

SIEGEL: It is pretty much an integral part of the concept of evolution that organisms either progress or retrogress. You either keep advancing, of course, we are not speaking of the individual now but of the type of organism, but ordinarily we would refer to the species, that life will either continue advancing, continually being shaped and pared down and moulded by the environment in which it exists and by its own inherent capabilities, to exist in that environment, or it will go downhill. I can't say that even this is a matter of proven biology, but it fits a great deal of our observations of nature so that advancement, progress, I should think is an integral part of our concept of evolution.

CARAS: Could I infer from what you said earlier that if an animal form of life arose on a planet and underwent an evolutionary sequence, that if this sequence were protracted enough, intelligence would inevitably be the result given enough billions of years?

SIEGEL: In spite of the fact that we recognise progress as an integral part of evolution, still the direction that that progress may take doesn't necessarily pass through a stage of intelligence. It would seem to me that the attainment of intelligence requires certain types of demands for survival. Demands that for example, an ordinary plant capable of producing a new generation by seed that could be easily propagated by, let us say, the wind or by birds but yet can't send out runners underground that will propagate itself as well. This is a very successful form of creature and perhaps it has different variations that can exist in Alpine and desert conditions. There are such versatile plants and original animals but with that kind of ability to reproduce itself in a range of environments and perhaps a seed that can lie dormant when the conditions aren't so good and wait until they are suitable in order to sprout. But with that kind of capability intelligence would be meaningless because what intelligence gives us is the ability to assess the situation of our environment and our relation to it and to either project from this or learn to anticipate or learn how to operate in the future on the basis of the experience and the assessments that we make in any one time and how to create from this where it would be meaningless for such an organism to develop this no matter how long it existed, so that evolution needn't always take the course, even if billions of years are involved, of going to intelligence.

SCENE TWO / TAKE TWO

SIEGEL: We can't have any facts at this point about life elsewhere. We can do experiments in the laboratory that say that, and they do say that whatever kind of stress we put Earthly life under, taking cognisance of the kinds of environments we do know exist on other planets, that whatever stress we put them under, there are very few kinds of challenge that really destroy life. They may impair it, they may destroy whole types of organisms but they cannot eliminate all life. Now this gives us confidence. This combined with our universal concept on the origin of life that once life originates nothing short of a sheer catastrophe such as the burning up of a planet somehow would really surely obliterate it. But I can only say that by conviction that the process of evolution is not something that begins suddenly with a cell or with the beginning of living matter and progresses until we get to some successful dominant form or forms of life, intelligent or otherwise. It is a process which begins essentially even below the level of the molecule, the level of the atom. I am not a physicist. I wouldn't venture to say that perhaps it begins at the sub-atomic level. Philosophically speaking there is a continuation of evolution from the simplest matter energy state to the astronomical level, to the cosmic level and then along this line of evolution or along the branching lines of evolution life is a state of matter and that whenever matter reaches a certain level of chemical and physical complexity, then it becomes possible for it to be alive, which means to assemble the components of its own substance, out of its environment, and thereby reproduce itself. The ability at the same time to react against its environment in such a way as to compensate for stresses that the environment produces, that if we have these elements together, the chemical matter, we have life or the beginnings of life and that this requires matter doesn't require anything specialised. Of course, I fully agree with the idea that our egotism has been a great factor in our thinking about life as a particular commodity of this planet in this solar system. It seems most unlikely to me that the chemical events as complicated as they are that have led to the formulation of the substance of life here can have happened only in one instance in the whole of cosmic space and time.

CARAS: Before, you were saying, I think it was that evolution, that life and matter, this is a continuance in evolution from inanimate to animate, evolution is unending. What follows life? If life then is a stage in the continuum, is it cyclic?

SIEGEL: I expressed a philosophy which says that life is a stage in the evolution of matter. If we examine this fully then it implies that we begin with some state, perhaps sub-atomic, and pass through its sub-stage, that which is living in the sense that life is familiar to us. But to be perfectly consistent, I would then have to say that there is a stage beyond living that is familiar to us in evolution, otherwise we say that evolution can reach an end but what could be beyond life as we know it? It is something which I, as an experimental biologist, cannot answer.

CARAS: Is it metaphysical?

SIEGEL: I suppose you could take off into the realm of metaphysics or theology or perhaps recognise some sort of an unending circular process and at both the cosmic level and at the biological level this concept is a very old one going back to the Greeks, the Phoenix concept. The concept of the universe burning itself up at intervals and a new one being born from the ashes, that is many, many centuries old.

CARAS: This kind of super-biology, super-bios. Would this be bios that follows? Would this seem to indicate, I know that there are no answers here, but as point of discussion is it possible then that the organism, the human organism if we consider the highest point of evolution so far on this planet, is this one day going to be free? Free itself of its present mortality and its present biological morphology and one day emerge as an intelligence, perhaps without a body? Is this kind of thing a point of discussion?

SIEGEL: One way of viewing this possible evolution of life beyond its present state is rather simple and direct. We already can recognise that there is a stage in the evolution of life beyond the cell, that is where the cell proliferates into many cells and these specialised in various activities and processes and we get a complicated organism. But then the organism in its own right is an individual just as the simple primordial cell was in its own right an individual. Now, the individual normally does not live as a solitary creature but in some sort of communal relationship with like and unlike other individuals be they plant, animal, or members of its same family or what-have-you. In a sense perhaps one way in which life is a solitary thing, we are each of us very much individuals, most of our feelings, most of our ideas and thoughts are all wrapped up within us, bottled up in fact in us. We cannot express most of what we feel and we are always on guard. We are still living back in the jungle in many ways. Perhaps the next stage is to open up man so that he becomes more of a planetary creature than an individual creature. This is not a political concept in the sense of international man but a biological concept of perhaps a development of processes, sensory processes, of communications processes which permit man to exist more fully, more wholly as part of this planet and from there on we can talk about going even beyond the limitations of the planet. It is hard to picture the way in which such a process of evolution would operate. We know, of course, that people have worked at extra-sensory perception and our projection of ideas, feelings and so on over a distance from one human being to another. The field is one that is still in doubt in the minds of many scientists and many laymen too but at least it has become worthy now of recognition as a field that can be studied and concerned about. Perhaps there is a mechanism here in much the same way as organisms that we know today generate electricity even in large packages, such as an electric eel. They turn part of their bodily energy into electricity. Others, such as the firefly, into visible light. Perhaps we can also turn part of our light processes and our energies into communications signals that others of our own kind can pick up. So maybe there is some medium for expanding our consciousness beyond the limits of our own skulls. This will be a logical next stage in evolution.

CARAS: If an organism arises on another planet, let us say, higher organisms arise on other planets, theoretical, their morphology is totally different from ours; their physiology totally different. It is a rattlesnake with a shell and the head of a whale but it has intelligence. It can communicate. It can recall. It can pass on history. It can reckon and it has consciousness of self. Do we call this thing man?

SCENE THREE / TAKE ONE

SIEGEL: When we think about the problem of encountering life elsewhere, the first level of concern is recognising that it is alive. Our next level of course is recognising the state of attainment of this life. Now, suppose we do encounter a life form that has by the evidence of its existence, its activities, whether these are spaceships or cities, or what-have-you, that has all the evidence of a highly developed intelligence that it could carry out all the processes that we could define ourselves as intelligence and most important of all, that it gives every evidence of having continuity in its own identity, self-awareness is another way of putting this. In such a case, how will we handle such an encounter? The way in which we handle this encounter will I think determine our whole relationship to this expansion of our own species into space in the future because we naturally tend to put a vague judgement on appearance, similarity and difference of appearance. We do this even among races. But what happens if such a creature, with all of its intelligence, perhaps superior to ours, is by our standards monstrous, ugly? It is really a matter of serious concern because we almost as a reflex shy away from things that are different. Suppose it looked like a spider, like a tarantula? This is for the sake of argument there would be one straightforward impulse that most humans would immediately fall to even if they were trained biologists, except for a few. That would be to step on it. And suppose it was of such a size that we couldn't step on it? It would be some problem of revulsion and it would take a very abstract and disinterested mind to adapt readily to try to communicate at its level of intelligence with something that looked like a tarantula. But we must recognise that if it gives all these evidences of intelligence, of self-awareness, of consciousness, then it is human in its standards as we are in our own. The word 'human' is our definition of our state of attainment and it must be the word we are willing to apply to any other creature that is in a comparable state.

CARAS: If an intelligence arose on another planet, on a planet that was older than our own, in a planetary system that was older than our own and the planet was say fifteen billions instead of five billions and the life form has had four million years of civilisation instead of the ten thousand we have had, and that is stretching it, what might their capabilities be? What might man's capabilities be in another four million years of intelligence? Of script and of mathematics?

SIEGEL: In trying to extend our thinking beyond the present state of evolution, we can look around us and see what is happening to various forms of life. For example, it is the belief of some biologists that certain species that have reached a sort of dead-end on the Earth as it is now can probably maintain themselves for a large number or more of years. Let me mention fossils, for some of these creatures like the virtue crab are so well adapted that until some major change comes that deprives them of their beach environment they will simply sit there, reasonably successful, not going any place but surviving. Other things. We can look back to the dinosaurs here: they became essentially unsuccessful, unviable within the physical limits of their environment but they ruled the Earth for tens of millions of years. It would be ridiculous for us to think that our own species does not have the chance too. Our own species, our own type of a primate creature that does not have the chance to become an evolutionary dominant for many, many times this small history that we have had so far.

CARAS: What might our capabilities be if our morphology stayed roughly the same?

SIEGEL: If we continue to progress along this ascending curve of knowledge and technology I would guess that well before we reach the traditional 'Buck Rogers' point, well before that, we will pass beyond any technology that we can even conceive of at the moment, that in terms of communications and terms of transportation, in terms of the whole of what makes the industry in our various respective world economies go are just going to be beyond anything we can conceive of. Such crude methods as the interchange of pieces of metal with arbitrary value put on them for currency. Things like this must be eliminated. Transportation must undergo tremendous changes and so on but to speak to the problem of a biological man it seems to me that long before we would think about even a multimillion year man I would say, giving man evolving for the next, evolving as an intelligent cultural animal, for the next few thousands of years that we will see the total elimination of disease. We will see the total elimination of what we consider genetic deficiencies. That we will see the opportunity for man to begin to be fitted into environments other than the general, typical human environments. That man is a creature perhaps breathing water will appear before that. We will be able to start engineering man as the centuries pass until the question of man in relationship to other planets will be partly solved by putting the right kind of Homo sapiens into the environment of the planet that we seek to colonise. If the population pressure, along with improving health and longevity, continues to increase then sooner or later there is no recourse but to colonise the other planets and this may determine a great deal of future human evolution.

CARAS: Since genetic engineering and engineering of form and shape and as you say of breathing water, which Cousteau[1] is actually working on, since these things are the products of human intelligence and human intelligence is arrived at through an evolutionary process, are these changes that will come? Are they evolutionary changes?

SIEGEL: Let us make sure of our definitions here of evolution. We talk about natural selection. The question to ask, perhaps quickly: is there any other kind? We breed animals and plants for our own agricultural purposes. We find certain facts in our canine breeds that we combine to produce a better hunting dog or what-have-you. Is this a perfectly natural selection by every standard? It is simply that man is the tool of natural selection and even when we engage in what might be called an operation of boot straps, the elevation of our own species, this is just as much a part of evolution as anything that happened before intelligence appeared on the scene.

---

1 Jacques Cousteau (1910-97), French oceanographer and undersea explorer. Caras is a little confused here. He is referring to Cousteau's Conshelf Saturation Dive Program in 1957 where men lived and worked underwater for a considerable period of time, but they certainly did not breathe it.

CARAS: Could you briefly describe what it would be like if we were on Mars and we walked outside the door? What would it be like? What would our senses tell us?

SIEGEL: Let us assume that we could have some direct hook-up so that even though we were humans this allows us to experience the environment of Mars directly and not clothed behind a spacesuit of some sort or protective garb. The first kind of thing that we would encounter is something we would find that is familiar to anybody who has travelled across the California desert, or any other desert area. This is the skin would begin to feel taut and parched.

SCENE FOUR / TAKE ONE

CARAS: We have omnipotent man who can survive anything in our theoretical situation stepping out of this room on to the surface of Mars. What would he feel? What would his, this sensation be?

SIEGEL: It is interesting to try to anticipate what it might be like to step on to the surface of Mars and really feel what that kind of environment might be. We might do this in theory by sort of inhabiting the body of a native Martian and then sort of plugging into the native Martian's senses. But I would think that if we used our own familiar Earthly sensations for what that environment might do to us, and step down on a Martian summer day, somewhere around the equator, one of the first sensations we would feel is extreme dryness of the environment. Very much the same thing that one feels, say in driving across the desert, spending just a day crossing the California desert. This is cracking of the lips, the skin tautens and so on and you would feel dryness immediately and, of course. in the mouth a parching sensation because it is so dry by Earthly standards (we don't have a desert that matches the dryness of Mars as a whole). On this day, of course, you would very rapidly feel the tingling sensation and the searing sensations in your skin that it had begun to undergo very unpleasant changes under the influence of the ultraviolet radiation of the Sun of which ninety per cent of the burning radiation is transmitted by the Martian atmosphere. There is no screen such as we have in our own atmosphere. Sunlight perhaps would be somewhat dimmer but not greatly so than it is on the Earth. The colour spectrum would be very much the same but your eyes wouldn't last very long without some sort of protection from the ultraviolet. Suppose you did manage to make it through a Martian day and the day in the summer at the equator might be in a comfortable temperature, all other problems excluded, you might have a nice 70 degrees day, and then as you see the Sun sinking, beyond the horizon, as its last rays disappear, you will suddenly feel cold. In a moment if you don't get out of that environment into suitable protection you are just going to be frozen solid and so this is the kind of situation that you will find on Mars.

CARAS: What do we hear? Do we hear wind? If we dropped a stone would it make the same kind of sound?

SIEGEL: The other senses such as auditory senses would be a little hard to judge. The atmosphere is so thin, one-hundredth or there-abouts of the density of the Earth's atmosphere, and its composition is different so that it is probably at the limits of human hearing if it isn't beyond the limits. Sound propagation in other words would be very, very rapid but the energy that could be carried, for example in terms of ordinary conversation, would be tremendously attenuated. I imagine one would have to converse either by direct contact like conduction or by electromagnetic means but not by ordinary conversation.

CARAS: Could there be a windstorm?

SIEGEL: This, of course, does not mean that the atmosphere is so thin that winds are non-existent. There is an atmosphere. There are windstorms. In fact one has to picture Mars as a place of intercontinental dust storms with winds at very, very high velocities moving across the face of Mars and in spite of the rarefied atmosphere having enough energy because of that velocity to lift up the surface dust masses until in some of the astronomical photographs, one sees two-thirds of the face of Mars obscured by yellow dust clouds. This is real wind – 200 mph is a figure that has been suggested for these winds. In other words, very much like the most severe winds on the Earth, and these would be not daily necessarily, but at least typical of Mars.

CARAS: What about the Moon? Again, a special kind of man who could survive anything? What would be his sensation of walking? What about the gravity on Mars?

SIEGEL: The other factor that should be brought to mind is of course gravitation. Mars has about four-tenths of the gravitation of the Earth, and so, of course, the buoyancy and lightness that this would give would be one of the most overwhelming sensations for any hypothetical man who could walk on the Mars surface unaided and, in fact, this would be important in terms of suited, protective suited individuals, because the burden of protective gear would be lessened thereby considerably. Now, if we go to the case of the Moon, it is really hard to project these sensations because we are beyond the place where even our hypothetical man is hard to put on the Moon, but there you would certainly have a world if intense light and shadow, intense extremes of light and dark, because without any atmosphere at all there is no scattering of light and therefore everything would be as dark as the depths of space, darker in fact, because there would be no stars. There would be the intense reflected light of the Sun. You would be in an utterly soundless world because of this lack of an atmosphere. Again, we'd have a world in which the temperature extremes would be even wider that they are on Mars but you would have to tolerate each of them, the hottest and the coldest, for about two weeks. You would have to adjust yourself to a day that is two weeks long and a night that is two weeks long, because these are the lunar periods of time.

CARAS: I want to go back before we finish to one thing we talked about before. You were saying that the genetic engineering, the biological engineering that can take place in the future are legitimately part of human evolution because they are part of the environment. Anyway, roughly this statement. What about machine intelligence? What about super-intelligent machines that the electronics men and the physicists predict? Is a machine with intellectual capabilities of man, if it ever arrives, is that also an extension of man? Is this also part of evolution or is this something totally different?

SIEGEL: In a sense. On the one end of the scale, the organic evolution, I picture the non-living mineral world that was the container for the primitive oceans in which life arose to have had a big part, the minerals that were lining the ocean basin, as having a big part in the origination of the organic matter itself that led to life. On the other end of the evolutionary scale there is another relationship of, we have to contemplate, living and non-living, in the sense now of man and machine where life originally was an organic material conditioned by the non-living world around it into the shape and form that it arose in. So, in turn, we are now creating a new world in which machine intelligence is playing a role. Now, some physicists and mathematicians predict of course super-intelligence in the machine level in the future. As a biologist I am not convinced that the lesser can create the greater, that we will ever create an intelligence, a machine intelligence, which is superior to our own. I am convinced that we can create machine intelligences that are faster than our own, that are more reliable than our own, but can we ever create an intelligence which is as capable of sound conclusions on the basis of incomplete and inadequate data as our own minds are? We do not operate like machines. We operate less like machines than some of our lower relatives such as the primates, lower primates. This is why a man is a much less reliable astronaut than a monkey, within limits, because a man pauses before he pushes buttons on command. A monkey doesn't. It does what it is told, a more machine-like type of existence. For generations, entomologists have likened insect behaviour to machine intelligence and if this is machine intelligence then it certainly is an entirely different entity than we picture as human intelligence and a far inferior one, except that it may be more accurate and more rapid, but can it ever be creative? If it cannot be creative then it can never excel man.

CARAS: Then you believe there will never be a machine that will be conscious of self?

SIEGEL: If there is, the point falls into another category entirely, that of human. It will no longer be a machine, but this is a matter of definition. I do not believe that a machine can be a machine and be conscious of itself. Now, whether man can create a new path of evolution in which he initiates processes which will lead to the development of some intelligence which is greater than his own is another matter, but can he construct at first hand that intelligence greater than his own? This is where my negativity reply to this question comes. Can he start an evolutionary clock just as this cosmos began? An evolutionary clock was turned on as it were and we had the progression from some primitive state of matter to the formation of galaxies and so on and eventually this comes down to the level of human evolution. Man might be the initiator of another chain of evolution which could ultimately excel him but it would excel him because it would be shaped by a long period of environmental process in its own right. Maybe a machine type of selection process could go on where in a robot jungle a more intelligent robot might have an advantage over a less intelligent robot and in that sense, yes, he could see the whole thing cast into a machine form taking place once again. But I think we will really have to go beyond this very unique ability of our own minds to create, and to create with very little information, to form judgements from faulty data, judgements that permit us to survive. We will go far, far beyond anything we can conceive of before any machine, any mechanical device can equal, much less excel, us.

B. F. Skinner

[Professor of Psychology, Harvard University]

SCENE ONE / TAKE ONE

CARAS: Doctor, twenty-five years ago if a graduate student in astronomy told his professor that he wanted to investigate the origin of life or do a paper on extraterrestrial life the possibility is he would have been told to leave the room. This Harlow Shapley has told us. Today it is a legitimate scientific pursuit. Although it is not precisely your discipline, do you have any thoughts on this? Do you know of anything that militates against the possibility of life on planets other than Earth?

SKINNER: Well, the possibility of life itself is hardly a psychological question. What form it would take, what a living organism would be doing in their relation to a special environment is intelligence. Life or mind are not special things, they are just ways we describe the kinds of organisms we are familiar with and these organisms have evolved because of the terrestrial environment. They have not only acquired certain characteristics of behaviour but they have acquired the ability to change very, very rapidly and we call this intelligence, or mind, or something like that, but all we know is that in this particular environment over a matter of millions of years organisms have developed with certain properties. In a different environment they would be very different and that would not mean they had a special form of intelligence. They would be structured a different way, behave in a different way.

CARAS: If life arises on other planets and the time period is protracted enough, can we assume that an evolutionary sequence would eventually take place or could life be static?

SKINNER: The actual evolutionary development necessary, that is rather inevitable, given certain conditions. That is something the biologist knows more about than I do. The final emergence of what we call intelligence in man does have very important psychological implications and the way in which we react to our environment because of the possession of special sense organs and ways of behaving and the ways in which we change our behaviour as we learn to adjust, that is a product of an environment as much as it is a part of the organism itself. The two things are closely related and a different physical environment would produce a different organism, and whether it would follow the same course would be very difficult to say.

CARAS: Astronomers have pointed out many times that there are planets with environments that are very, very similar to Earth, have the same gasses, the same water contents, the same gravitational field. Can we assume that if life would arise on one of these planets and go into an evolutionary march of some kind that intelligence would inevitably be the result of such an evolution in an environment similar to Earth?

SKINNER: Well, intelligence would necessarily develop not as a stuff, not as a special mind stuff, not anything of that kind, but whenever an organism changes so that it can react more effectively to the world around it, it has made some little step in the direction of what we call intelligence. This is an acquired way of reacting to an environment. It has acquired ways of changing its behaviour with respect to specific environments and that is rudimentary intelligence. Now, whether what we regard as the intelligence of man would emerge depends upon whether an organism something like man would emerge in something like the same environment, but we have no way of knowing how far another world would permit this development.

CARAS: How closely tied in are intelligence as we understand it and morphology? Could, for instance, a morphological shape of the shape of another creature of another planet be totally different from ours with an intelligence that was generically of our type?

SKINNER: Well, the biologists seem to feel that you would generally have to have something like the same chemistry, water, carbon and so on, and if that led to organisms something like ours then the same kind of intelligence would develop in the sense that the possibilities of altering the organism by selection mutation would be relatively the same and, I suppose, in that case the behaviour of it, I would prefer to talk about the behaviour of the other organism rather than the intelligence, and how an organism not like man would behave is the issue. Would he have verbal behaviour and if so would he be vocal? Or it might be gestures or something of that kind, but I dislike the view that somehow or other there is a special kind of thought or a special intelligence with which we have to communicate. It would be a question of whether or not the verbal stimuli we create would have an effect on such an organism and whether the verbal stimuli it can generate would be interpreted by us correctly.

CARAS: Civilisation on Earth, if we stretch it to the breaking point, is ten thousand years old. There are planets that are very much older than Earth and presumably if they have life, the life is very much older than ours. It has had longer to evolve. It is conceivable that there is a civilisation somewhere, let us say, that is four million ten thousand years old instead of ten thousand years old. What might the capabilities of an intelligent organism be that had four million ten thousand years of script language, territorial structure?

SKINNER: If you assume that the organism itself has not changed much in that four million years as we do assume, that man has not changed substantially in ten thousand years, if it is the same man then the change would be in the direction of a much more elaborate culture. He would have learned much better ways of harnessing his own energies and developing his own potentialities and in the field of verbal behaviour, of science, art, literature, all of this would have continued pretty much along the same track, but I don't believe we can possibly guess at what it would be like. We can't even predict what our own is going to be like ten to fifteen years from now very successfully.

CARAS: Would you care to extrapolate on some of the major changes that man will undergo? Man's social structures? His intelligence? The so-called intelligence explosion? What is coming for man in this environment in the immediate future?

SKINNER: Well, again, intelligence is not a thing that is going to change. Man is going to develop more and more subtle forms of behaviour, presumably in the field of science and genetics. He may actually begin to change himself as an organism but we will dismiss that. I know nothing about what could happen there and what we would want to happen, but we can certainly make vast changes in our culture. Better ways of teaching, better ways of harnessing productive energies, better ways of building sensitivity to art and so on. More secure governments, better justice and so on, all of this we can predict. We can see the direction it will take in a not too distant future and actually make some substantial proposals with some chance of success.

CARAS: There has been a certain amount of hysterical reaction on the subject of extraterrestrial life, evidence of UFOs. The most hysterical of these seem to be people who have been to Venus, Mars, Jupiter, as guests. They generally describe the visitors who take them away as having aquiline noses, long white beards and long white robes. What does this tell us?

SKINNER: The speculation about what extraterrestrial beings look like and stories of those who have claimed to have seen them, that is just a current form of psychosis. Every period has its own forms… the kinds of visions you see, the kinds of people who are whispering in your ear and so on. They are fashions and they come and go. This suggests a little to me the type of organism which appears to have been encountered, suggests a sort of blend of an angel for example with a robe or something like that and a facial type which is clearly in our own species but distant enough not to be one of us.

CARAS: I have asked this question both of a priest and of a rabbi and of other scientists and we don't get very often a very clear defined answer. Mankind for a very long time has referred to a place, in various languages, in various semantic contexts, of Heaven as an abode of a Prime Being. Heaven is described…

SCENE TWO / TAKE ONE

CARAS: Doctor Skinner, cultures for a very long time have pointed to Heaven as an abode of a supernatural being. Heaven is not in a tree, it is not in a rock and it is not in the middle of the Earth. It is always 'up'. Some people find this significant. What is the association with 'up', 'Heaven', 'space', and God?

SKINNER: I suppose Heaven has always been located 'up' and Hell 'down', because of the seasons and various natural events. Things come down from Heaven, rain, wind, lightning and so on. In winter everything goes underground and this is the old Greek myth of the winter and the spring. Things do come up again but only after the sun has come down, and so on. I suspect that Heaven in that sense is a reference to natural forces which happen to be above us. Of course, it is only recently that we have been able to discover what is above us, very far up.

CARAS: What are the chances of cultural shock and what might the nature of cultural shock be if it should be announced tomorrow, or next year, or ten years from now that we had definitely received intelligent, intelligible radio signals from outer space of an artificial origin? A mathematical code, for instance?

SKINNER: I don't believe many people would be greatly startled to discover that there are other intelligent beings in the universe who had advanced to the point at which they had discovered mathematical formulae and could send us some universal constants so that we would recognise it as the same kind of evolution of an organism and of a culture… The shock to our culture would not be, I suppose, as important as the shock of Europe discovering China, Marco Polo. That was an astonishing thing and it had an effect because you could get there fairly soon and things came back and you saw for the first time. Really, at that time in history, Western civilisation had a culture very much unlike our own. Now, of course, mathematical formulae being sent through space would give us very little information about the type of organism sending it. We could imagine all sorts of different physiognomy and so on but it would have to be a development which had led to the same kind of verbal and symbolic activities as our own.

CARAS: People will be talking now endlessly about the intelligence explosion, about the development of machine intelligence. There are many arguments whether man's space probes are justifiable enough. We should not send robot machines. There is a lot of work in this area. Where is machine intelligence taking us? When are we going to have a machine that is more intelligent than man, the so-called super-intelligent machine?

SKINNER: I am not sure that machines will become more intelligent than man. Machines are used. We have had labour-saving devices for a very long time that replace our muscles. We have sensing devices which replace eyes and ears, do a much better job. There is no reason why we can't have machines which will bring these two together and integrate them in a way in which the rest of the organism has been doing in the past, but it will be supplementary. Man uses the machines to move things. He uses sensing devices to see and he will use computers and other devices to arrive at formulation which will make him more effective. I don't believe there is anything called 'mind' as such or intelligence and I don't think it can be possessed by a machine. A machine is simply a product of human ingenuity which man uses to behave. It is an instrument and I cannot imagine that it would be in any sense superior to man; that will be used by him for presumably achievements which are way out of range of anything in the past, I am sure of that. But they will be human achievements.

CARAS: You don't see any possibility of man being superseded in vast areas by machines, intellectual areas?

SKINNER: I don't think man is going to be replaced as a thinker by machines. He will be improved as a thinker by them. He is not replaced as a mover by the big earth-moving devices he uses to dig holes and so on. He is using them and he is not replaced by the telescope, microscope, electronic microscope and so on. He eventually sees what those things give him to see. It is an extension of his own body and a computer is an extension of his own body that simply amplifies human behaviour, make it much more effective but it will be human.

CARAS: As machines become more and more complicated is there any possibility that they will develop personalities and personality disorders?

SKINNER: I think the notion of a machine having a nervous breakdown is fanciful. They break down of course, but if you leave out the word 'nervous', because it is not exactly a nervous system, it is an electronic breakdown or something of that sort, then the thing becomes trivial. Of course, they have breakdowns, but then that is not parallel with the breakdown of the human organism, because it isn't a human organism. Now whether or not man, by using these devices, will get himself into spots where he breaks down along with the machine is another matter. I suppose in a sense man may discover too much through, let us say, some fantastic new sensing device for him to handle. In that case he will get too much information and the whole thing might be too much for him, but it is the man who is breaking down and not the sensing device.

CARAS: Doctor Skinner, the machine that is being developed and which will be far more complicated than ourselves. Some people believe it will be an electronic device like a computer. Dr Freeman Dyson of Princeton told me he did not think it would be, he thought electronics was severely limited and that it would be a biological machine, a bios machine. Do you have any thoughts about this?

SKINNER: The beautiful way in which the brain works in the very limited space compared with even the most miniaturised computers would suggest that there are better things than electronic circuits. There are sticky states to the solid state systems which may very well be developed eventually which would make possible even more elaborate networks that are certainly in a much smaller space. At least that is what the human organism has already done in evolving its own brain. Whether this is the direction that will be practical to take, thinking machines are completely a matter of technology, what is available, what can be done. I suppose that our increasing ability to push molecules around and build them and put them together and so on, may very well lead to something. It may be that a solid state system is on its way to a sticky state.

CARAS: The human body is undergoing evolution. It has not arrived at a final form yet, there is no reason to believe. Is the mind similarly evolving?

SKINNER: I do not believe in the evolution of 'mind' as such. That sort of thing seems to me to be nonsense. What are evolved is a human organism and human culture and the achievements of mind. The so-called products of mind are simply the products of that organism, given that culture and our future development in the world will not be a refinement of any mental processes, it will be a refinement of our own organism as such by further evolution or by deliberate genetic change and particularly the refinement of our culture in the devices in that culture. Naturally computers are part of our culture, just as much as use of a tool or a stone or the flint scraper. A computer is the same kind of thing and it vastly increases man's effectiveness.

I don't believe there has been a special kind of thing called intelligence or mind which has emerged at any point and is now moving in the direction of getting better or more sensitive. It is the human organism we are always concerned with, what it can do as a biological system given a culture as an environment.

CARAS: Very briefly, Doctor, what percentage of our brain do we use? What percentage of our intellectual capacity do we use?

SKINNER: I do not believe at the present time that our educational systems and our environment in general lead us to behave… Oh, I wouldn't want to put a finger on it, but let us guess it. Ten per cent of what we might be capable of under a better-designed culture. And it is up to us to design a culture which will improve the human organism and make it more effective. I am sure that could be done. I would say ten per cent might be a generous estimate!

Fred L. Whipple

[Director of the Smithsonian Astrophysical Observatory; Professor of Astronomy, Harvard University]

6 MAY 1966 / SCENE ONE / TAKE ONE

CARAS: Doctor, is there anything in our knowledge, within the framework of our knowledge, that militates against there being life on planets other than Earth?

WHIPPLE: I should say that all the evidence we have includes the high probability that there is life on some planets in the universe and very likely on huge numbers of planets.

CARAS: What evidence, not necessarily empirical, but what would lead people to believe this? What leads scientists? They seem to mostly believe this today.

WHIPPLE: As we study the problems of the evolution of planets we find that the type of planet that we live on appears to be quite possible, quite general in the universe, and we have direct evidence of some of the larger ones such as Jupiter type planets and this leads to the deduction that conditions similar to these on our Earth must be prevalent in a great many planets. Now, making that assumption then one only has to turn to the biological evidence that suggests very strongly that life can develop where conditions are suitable, so that with these two bits of evidence we can't say positively that life exists but we can say that the possibility is strong.

CARAS: Do we have any grounds, any basis for drawing assumptions as to what this life might be like?

WHIPPLE: The problem of what the life might be like – we have a lot of fundamental problems to answer. I would say that the biologists are the ones who should attempt to answer this as to what life should be like and they are very careful. I don't think I should put words in the biologists' mouths.

CARAS: Sir, what are your theories or thoughts about the origin of life on this Earth? Or, for that matter, any planet?

WHIPPLE: Well, as a working assumption I assume that life originated as a natural result of the basic laws of chemistry and physics and I think that all the evidence of recent years points to the likelihood that it develops directly as a consequence of the laws of matter.

CARAS: Do you subscribe at all to the panspermia theory? Do you give any credence at all to the panspermia theory?

WHIPPLE: This question of spores surviving interplanetary or interstellar travel does not appeal to me a bit. In terms of my knowledge of interplanetary material we have strong evidence that etching of small bodies in space is appreciable so that if you are going to start life on some distant planet and then pass the cells along by matter that is blown out of a planet, or moved from a planet to another one, or between stars, you have an extremely serious problem of protecting these from complete destruction in the passage.

CARAS: Are planets the only celestial bodies that would appear to be possible, or probable residences for life forms?

WHIPPLE: To visualise life on any other object in space or in any other part of space than on a planet, then one has to postulate a completely different type of life than anything we know about and, of course, in science fiction that has been dealt with. Gaseous spheres with some signs of homogeneity and intrinsic coherence. But again, from one's knowledge of physics, one knows in gasses you have the entropy laws and very soon there is no organisation left. So I think it is impossible to build up organisations except in a liquid form or conceivably in some solid form and if you build life in a solid form you still must have something like a planet.

CARAS: Would any of the asteroids be large enough?

WHIPPLE: I would say that the chance of life beyond the extremely primitive forms of perhaps cells or viruses is something that would be quite impossible on the asteroids because of their small size.

CARAS: Could we assume that when life arises on a planet that it will automatically, if the time is protracted enough, undergo an evolutionary cycle of some kind or could it be static?

WHIPPLE: This question of the actual evolution starting from the atoms, moving up to complex molecules through to life, is a biological question and whether it should stop at some point and reach a point of stasis, I can't answer. I presume it might if conditions were possibly too uniform. I presume that you might halt the process almost at any point if the conditions were unchanging and there was very little variation in the temperature, pressure, chemistry and so forth.

CARAS: If an evolutionary march is underway could we assume that intelligence would be the inevitable result of evolution, or is it not necessarily so?

WHIPPLE: This question of whether biological evolution must lead to intelligence is a question of probably assuming an answer before we ask the question! I have only an opinion or a prejudice that biological evolution will lead to higher and higher life forms, but I don't think there is any sound basis for assuming that that could be true.

CARAS: Doctor, there has been a lot of nonsense about UFOs of course and about visitations and people who have gone to Venus as contacts. Is there any evidence at all, empirical or otherwise, that the Earth has ever been visited or surveyed or touched upon in any way whatsoever, by any extraterrestrial life form? Is there any reason to think this might be so?

WHIPPLE: Well! With regard to the UFOs and any possible evidence for extraterrestrial life having visited the planet, I consider this a religious subject, a cult, and is not one subject to scientific answer in view of the fact that we have no positive evidence. Therefore, I prefer not to discuss the UFO problem.

CARAS: Skipping UFOs completely, and I agree with you of course, we should not dwell on it. I know there is no positive evidence. I am talking now about the possibility of the planet ever being visited in the past. I am not talking about flying saucers! Is there any kind of evidence?

WHIPPLE: With regard to the possibility of the Earth having been visited in the past, I would say that the possibility exists, but I know of no evidence to suggest it.

SCENE TWO / TAKE ONE

CARAS: How shall we learn about extraterrestrial life first? How will the news come to us?

WHIPPLE: With regard to learning about extraterrestrial life, of course, our first move is to go to the planets and study them as we are now planning and have been carrying out some preliminary studies and NASA, the National Aeronautics and Space Administration programme. That's the first move. Now, if we wish to consider the possibility of proving that there are intelligent living beings on other planets, it seems unlikely that this could happen in our solar system. The evidence that has accumulated with regard to the more likely planets, therefore, they will have to live on planets revolving about stars, and all stars are physically very distant. Now, there is one possibility of just listening for them. This has generally assumed that intelligent creatures somewhere are so anxious to make their presence known or to communicate that they spend huge fractions of their resources in powerful transmitters that could transmit for tens, hundreds or thousands of years into space in the vague hope that somebody will answer them. Now, this type of communication I think is highly unlikely. I would suspect that although all cultures might not be like ours, there has to be some practical limitations for survival and they are not going to spend all their resources in this type of effort. But what might happen is that one will find a planetary system in which there are several inhabitable planets, two or more at least, and that then they will establish trade and communication. If we could locate such a system in which we lay very nearly on the plane of their mutual orbits or somewhere near that plane, then there might be a chance that we would pick up communication signals when two or more planets were properly aligned to send a signal in our direction. Now, whether this is worth listening for or not is a real question in my mind. Whether the chances are high enough and whether we can ever detect such communication in such a reasonable period of time is uncertain.

CARAS: Surely, twenty-five years ago in astronomy, astronomers were not seriously anticipating men going out into space for direct observation? Dr Shapley was saying that people today are studying seriously in depth subjects which, if proposed to him by a graduate student twenty-five years ago, would have felt the toe of his shoe. How has astronomy changed? Where are the big changes in attitude and ideas?

WHIPPLE: This question of the endeavour to send men into space and the changes of attitude that have taken place in astronomy represent a practical situation. Now, twenty-five years ago, I certainly for one and a great many other friends of mine who were interested in the possibilities of life in other parts of the universe put a great deal of thought into the possibility of man going into space and I remember the exact day when I became certain that this would happen. This was back in 1944, which is now twenty-two years ago, when the first V2 was launched, albeit on a very hapless mission. Nevertheless, at that time my long anticipated thought that we would go into space became in my mind a certain reality for the future. I don't know how many other people thought in the same fashion. I would say for most of my life the expectation that man would go into space had been very real in the minds of many scientists, many astronomers. On the other hand, this had no practical aspect of application at the time when the technical possibilities had not been developed. There was no chance of exploiting a technology then. Now, since the V2 in 1944 it has been quite clear that it is technically possible to do this and the question has been just how and how soon. So, since that time I think a lot of scientists and some astronomers have contemplated very seriously this problem of what man should do in space and what we should do with space vehicles.

CARAS: 1944. Did you think of it as a working tool in a quarter of a century?

WHIPPLE: In 1944 I knew that man was going into space. In my own mind. I was certain. In 1946 I did my first research paper on the subject, which was about as far as I could go at that time. I invented the meteor bumper[1] then as a means of protecting space vehicles from meteoric impact in space. So at that time I wasn't very sure in my mind how long it would take; but I certainly measured it in a few decades and not in terms of a hundred years or any such period of time as that.

CARAS: At a luncheon last Tuesday, a week ago yesterday, Dr von Braun was holding forth pre-eminently on the significance of man's space flight as opposed to unmanned space probes. Already this week two physicists at MIT held forth just as strongly that it was absolute nonsense. It was a great show, great drama, and it was a wonderful publicity stunt but there was no justification at all for sending man into space. All could be done by remote vehicles and this argument has been going on for a long time with some very impressive credentials on both sides.

WHIPPLE: This problem of man in space versus instruments in space is, of course, a difficult one. The question really revolves about the practical costs and the resources that one can invest in such matters. My own attitude has been that it is possible to send man into space and that any culture or country, or even person for that matter, who sees a challenge that he might attempt to meet new frontiers of any sort who doesn't accept the challenge and attempt to cross the frontier is going to be decadent. I think we have to consider and definitely plan for man in space as well as for unmanned instrumentation in space.

CARAS: Does this include deep space as well as the space between our own planets in this one solar system?

WHIPPLE: With regard to sending man into deep space versus instruments into deep space, again the natural resources are the more important factor. I think I would be rather satisfied, for example in studying Jupiter, simply to send unmanned instruments. On the other hand, satellites of Jupiter could be very exciting. The landing of a man on the satellites of Jupiter could lead to some very exciting results. And then if one carries his imagination a bit farther into the future and visualises, from our point of view, almost infinite

---

1 Also known as the Whipple Shield. He invented it in 1946. Aside from his many outstanding achievements in astronomy, Whipple also came up with the method of using aluminium chaff as a countermeasure to enemy radar in World War II.

energy sources, stronger materials, better technology, it might be that the man down inside the atmosphere of Jupiter under high gravity, and so forth, could conceivably make discoveries that the instruments would miss. Also, in some cases, such as that, it may be extremely difficult getting information back by any radio communication device. It will be necessary to return something. You might as well try to send a man too for what he can add to the observations.

SCENE THREE / TAKE ONE

CARAS: Doctor, you said before that life we find on Mars is going to be very primitive. Are we going to find anything in the solar system, our own solar system, that will ever justify the sending forth of a palaeontologist or an archaeologist?

WHIPPLE: When it comes to the problem of the complexity of life that we may or may not find on some other terrestrial planets, I speak with great conservatism. The older I get the more I find that negative statements tend to be dangerous at least. It is quite true that the evidence say, on Mars, suggests that you wouldn't expect very large organisms and this is very probably true. When one considers possibilities we should speak very carefully because it is quite possible under the surface, coming up occasionally, there might be sizeable living organisms and even conceivably – I think that extremely unlikely – intelligent organisms, and we are not sure enough in the case of Venus to state with absolute precision that somewhere there isn't a habitable area. So, I would say there is a good chance that a palaeontologist might be, no, there is only some chance, that a palaeontologist might be needed somewhere in the solar system.

CARAS: When we go forth, where will we colonise and how will we do it?

WHIPPLE: Colonisation in the solar system of course at the present time is in the dream state, but I would say that if large power sources such as the fusion process become readily controllable and generally available, we could think very seriously of supplying an atmosphere for Mars and in the case of Venus I suspect that may be more difficult to do although there has been some suggestion of altering the atmosphere by biological techniques, possibly other chemical techniques. One should not rule that out in the long run. I rather feel that the Moon is one of the bodies that is less likely to become, shall we say, economically feasible for colonisation because of the inability I think of man to supply an atmosphere that will remain long enough to be economically feasible. The motions of the molecules will encourage them to escape from the Moon so rapidly that I suspect that that is not going to be a very good object for which one could hope to supply an atmosphere.

CARAS: In the scheme of things is Earth an old planet or a new planet? Is it probably a large planet? Is it probably a small planet, thinking in terms well beyond our own solar system?

WHIPPLE: The position of the Earth with regard to other possible planets throughout the universe is not very well established. I would say that compared to planets as a whole it is a relatively small one, in terms that is, if one confines oneself to planets that are large enough to keep atmospheres. In fact, as we see when we go to a planet roughly half the dimension of the Earth. In the case of Mars we have got the ability to hold an extremely thin atmosphere. If one generalises the definition to include the asteroids or minor planets, because they go down to rock size and in that case the Earth would be quite large.

CARAS: Is Mars a young planet that is developing or is it an old planet that is dying?

WHIPPLE: The age of the Earth or Mars is fairly clearly defined now. I think it is fair to say that both of them are rather old planets. By old, I mean they have existed for a long period of time and the transformations in the interior are fairly complete. I don't think much change can take place there and generally the changes on the outside are going to be of the dissipative character as the time goes on. So you might say that the Earth is perhaps middle-aged and Mars is rather old, because a smaller planet will develop more rapidly than a larger one.

CARAS: An eminent scientist told me last week that there were no conceivable grounds for believing that we will ever be able to project ourselves or anything faster than the speed of light. An equally eminent scientist told me earlier this week that there is absolutely no reason for believing that we could not do it.

WHIPPLE: Can we travel faster than light? That question, of course, is the most difficult one to answer with any assurance. No one can give a positive answer with any assurance. No one can give a positive answer. I think it is fair to say though on the basis of all the physics that we know today that one must conclude that it is impossible, but that doesn't preclude an extremely remote possibility that we have erred in some way as we have in the past in predicting what is possible and what is impossible.

CARAS: Is there anything you would like to add in this general theme?

WHIPPLE: I don't think so because my point of view on people travelling through space is rather negative, based upon the little evidence, the negative evidence, we have of ever having done so. Maybe I might make this statement: I have been highly suspicious of travel from one planet to another, at least through interstellar space or unless two stars happen to be relatively close together in passing. The reason for this is that it is an extremely expensive process in terms of physical resources, and demands activity of such magnitude that I really question that it happens very frequently. I am sure attempts of this sort must have been made in galactic time and certainly will be made. The chance of it becoming permanent, or frequent since any one planet is visited often, seems to me very remote.

CARAS: The fact that you say you feel sure this has happened some time in galactic time expresses a very firm belief in extraterrestrial intelligence of high sophistication.

WHIPPLE: I would say that I do have a firm belief in extra-galactic intelligence, cultures and life forms in the universe. This, of course, is a type of faith I suppose.

CARAS: How do you suppose we will react to them when we finally make contact, radio or otherwise? Will it be a big cultural shock?

WHIPPLE: My guess is that when we do, if we survive on this planet for enough hundreds of millions of years and meet extraterrestrial beings, that probably the differences in evolutionary stage or in cultural objectives will be so great that even an intelligible communication would probably be very difficult.

CARAS: But quite accomplishable?

WHIPPLE: Barely accomplishable. But I don't think there is much likelihood that we will really understand each other. My real concern is nuclear warfare, which may wipe out cultures throughout the universe.

Appendix I

A Theological Take: God in *2001: A Space Odyssey*

by Dr John Braverman SJ

THE FILM *2001: A Space Odyssey* and the *Playboy* interview with its director, Stanley Kubrick, are the subjects of this essay. I write from the perspective of a Jesuit scientist, which means that I am curious about the cosmos we humans inhabit, and the philosophical and theological implications of that cosmos.

Here I reflect on the discovery of intelligent extraterrestrial life in the film *2001*, and Kubrick's statement that 'the God concept is at the heart of *2001*.' I argue that Kubrick's work is far more traditional than he says, although it is nonetheless forward-looking.

The interview Kubrick gave to *Playboy* in September 1968 is a very helpful window into *2001*. In this interview, Kubrick resists interpreting his film. This leaves the burden of interpretation on the viewer, although Kubrick wanted the film to be 'experienced,' rather than interpreted, through multiple viewings and extensive study. Still, after multiple prods by the interviewer, Kubrick does make a brief interpretive statement:

> I will say that the God concept is at the heart of *2001* – but not any traditional, anthropomorphic image of God. I don't believe in any of Earth's monotheistic religions, but I do believe that one can construct an intriguing scientific definition of God, once you accept the fact that there are approximately 100 billion stars in our galaxy alone…

Kubrick proceeds with the argument – now familiar to the general public – that the existence of intelligent life elsewhere in the universe is highly probable due to the huge number of stars and orbiting planets. He extends the argument one step beyond just saying that intelligent life exists, suggesting that odds are that at least some of that life is advanced far beyond us:

> Can you imagine the evolutionary development that much older life forms have

> taken? They may have progressed from biological species, which are fragile shells for the mind at best, into immortal machine entities – and then, over innumerable eons, they could emerge from the chrysalis of matter transformed into beings of pure energy and spirit. Their potentialities would be limitless and their intelligence ungraspable by humans.

The beings in *2001*, then, who planted the monolith on prehistoric Earth and on the Moon and who greeted astronaut David Bowman near Jupiter, might be these beings of pure energy and spirit. Relative to humans, these extraterrestrial beings are gods. As Kubrick notes in the same interview, 'They would possess the twin attributes of all deities – omniscience and omnipotence.'

Kubrick's 'God concept' as portrayed in the film is a little more than omniscience and omnipotence. The aliens are in a serious relationship with humans. They have an interest in humans from the very beginning. The prehistoric humans learn to use a bone as a tool, in close proximity to the first monolith. The second monolith is planted in the Moon as a kind of 'sentinel' (to use the title of Arthur C. Clark's short story), waiting the moment when humans have reached a certain level of technical skill and curiosity. This monolith emits a signal aimed at Jupiter when discovered by humans. If the alien beings, though, were truly omniscient, they would not need this complex mechanism to signal them at the appropriate moment. So, the primary function must be to tell humans where to go next to meet the aliens in person.

The aliens are interested in human progress. The *Playboy* interviewer seems to realize this is unlikely for omniscient, omnipotent beings, and so he asks Kubrick why the aliens would be interested in humans? Although Kubrick would not speculate as to the motives, the simple answer is that the aliens are not merely omnipotent and omniscient. True, a distant, unconcerned god is the 'God concept.' But historically, the concept comes from Aristotle and other philosophers who focused chiefly on the intellectual power of God. They did not treat another God-trait often mentioned throughout the ages by traditional religions: love.

Let me express this in a different way. I agree that the 'God concept' is present in the film *2001*. Minimally, it is a particular God concept, which is known as the 'God of the philosophers.' That conceptual God is not the personal, loving God of intimate connection familiar to Isaac, Abraham, and Jesus. On the other hand, this otherwise distant alien race of omniscience and omnipotence did decide to make contact with humans, and they seemed to have a long-term commitment to follow human progress. They did not interfere in human development, just as a loving parent lets its child grow up to make his or her own choices. In this way the aliens did display a kind of love – the kind which lets humans be humans, and which only rarely makes subtle prods. The radio signal calls us forth. So it is not merely the 'God concept.'

To return to Stanley Kubrick's comments in the *Playboy* interview, he suggests a 'scientific definition' of God, offering a very insightful theology. First, the logical steps described by Kubrick are more than a definition: they are the proof for the existence of the god-like intelligent alien beings. Proofs for the existence of God often follow similar logical lines, and comprise the field of natural theology. The most famous of those proofs appear in the *Summa Theologiae* of St. Thomas Aquinas. Traditional proofs use deductive reasoning rather than Kubrick's probabilistic or statistical reasoning. But they parallel in one dramatic way: use of information about the natural world to infer the existence of higher intelligent beings. In that sense, Kubrick's method is perhaps more appropriately labeled 'empirical' rather than 'scientific.'

Although natural theology has fallen on hard times, the tradition of seeking understanding of God's creation remains a strong motivation for Jesuit scholarship. Part of this interest is exemplified in various Jesuit astronomers, including Father Francis J. Heyden, S.J., a professor of astronomy at Georgetown University, whom Stanley Kubrick interviewed during the making of *2001*.

Another striking 'Jesuit connection' is the location of the monolith: the Clavius zone on the surface of the moon. Christopher Clavius is one of 35 Jesuits who have moon craters and other formations named after them. Part of that honor is due to the extensive astronomy work done by Jesuits since the early days of that field.

I would like to comment on the very last scene in which the star child appears. This is apparently astronaut David Bowman reincarnated as a fetus, at once fresh and staring serenely with new-found wisdom. One could focus on that wisdom and knowledge gained perhaps from humanity's third encounter with the alien monolith. Yet that does not make perfect sense if Dave is now a baby and has to do it all over again! I do not assert that Kubrick chose the fetus image for this reason, but the choice strikes me as eminently consistent with the god-like aliens who love humanity as humanity. Again, the theology goes beyond the 'God concept.'

The encounter with alien intelligent life does not wreak havoc in human self-understanding. Ironically, it pointed humans back to themselves. In his odyssey, Dave disconnects HAL, making this voyage a human achievement. Dave's reappearance as a human fetus, the star child, invites another round of human development. Maybe the aliens are displaying this fetus, intrigued by the human story. The aliens are humanists in that they are not going to tease us or short-circuit humanity's long development. This is a very curious turn of events in light of Kubrick's desire to avoid another 'anthropomorphic image of God' in the film.

Kubrick may not 'believe in any of Earth's monotheistic religions,' but he may have used some traditional theological methods in his interview with *Playboy* and in the film *2001*. My comments here represent only a brief reflection on the topic of God and *2001*. Much more could be said about this fascinating movie and its intriguing director.

Chicago, 2003

[John M. Braverman, S.J., Ph.D., is a Jesuit and a visiting assistant professor in the Department of Biology at Georgetown University.]

The Monolith suspended by wires from a crane on the back lot of MGM studios

Appendix II

Other Scientists Who Were Approached

THERE WERE MANY reasons for scientists not appearing, ranging from pressure of work or holiday commitments, through to their position not allowing them to seemingly 'endorse' a commercial film venture.

Harold C. Urey, famous for his work with Stanley L. Miller on the origin of life, responded to Roger Caras' invitation as follows from UC San Diego, 7 February 1966:

> It would seem to me improbable that one would find an artefact on the moon of extraterrestrial intelligence and I would not like to introduce a film that presents this point of view.[1]

So there! But he held out hope:

> I confidently believe that there is no reasonable possibility that intelligent life existed on Mars or Venus. The most that I have hoped for is that there might be some evidence of primitive life on these planets, or at least that such primitive life existed some time in the past.

One letter of refusal rather highlights the view of ETI held by older, more conservative scientists and it is worth quoting to show the regard in which it was held in 1966. This is Dr. Sidney W. Fox at the Institute of Molecular Evolution, University of Miami, writing to Caras as of 4 February 1966:

> In response to your inquiry of 28 January, I am not the one to discuss for you

1] We were not alone. Urey turned down a request to write a blurb for the I. S. Shklovskii/Carl Sagan study, *Intelligent Life in the Universe* (1966), on the grounds that 'he didn't think a book on the subject of extraterrestrial life could really have any value, given how little we know about the subject.' Quoted in Keay Davidson, *Carl Sagan: A Life* (New York: John Wiley, 1999), p. 198.

> extraterrestrial intelligence, especially since I am sharply critical of detailed treatment of entities of which the reality has yet to be established.

And further:

> For your specific request, however, I regard the subject matter as possibly factual, but probably not so in the usual nonevolutionary [*sic*] context of (humanoid) intelligence. I do regard it as close to fantasy. Having listened this week to an exposition of the search for intelligent extraterrestrial life, I know that you can find some who will make your point, but I am sure that the vast majority of realistic scientists will not be amenable, and I would deplore any attempt to make any intellectual exercise appear to hold more promise than it does.

The following is a list of those who were approached, based on surviving paperwork. There may have been others. In alphabetical order of surname:

Hannes Alfven, Royal Institute of Technology, Stockholm
R. M. Bracewell, Radio Astronomy Institute, Stamford
Wernher von Braun, NASA, Huntsville, Alabama
Michael H. Briggs, Analytical Laboratories Ltd, Wiltshire
Melvin Calvin, University of California, Berkeley
A. G. W. Cameron, Goddard Space Flight Center
R. R. J. Chaffee, Space Sciences Research Center, University of Missouri
A. Dauvillier, Observatoire du Pic du Midi, France
Sidney W. Fox, Institute of Molecular Evolution, University of Miami
Owen Gingerich, Smithsonian Observatory, Massachusetts
John H. Glenn, NASA astronaut
S. von Hoerner, Green Bank Observatory, West Virginia
Su-Shu Huang, Dearborn Observatory, Northwestern University, Illinois
Robert Jastrow, Goddard Space Flight Center
Louis Leakey, anthropologist, London
Joshua Lederberg, Dept. of Genetics, Stamford University
John C. Lilley, Communications Research Institute, Florida
Stanley L. Miller, University of Chicago
J. P. T. Pearman, University of Rochester
N. W. Pirie, Rothamstead Experimental Station, England
Edward M. Purcell, Dept. of Physics, Harvard University
I. S. Shklovskii, Astronomical Institute, Moscow
W. M. Sinton, Lowell Observatory, Arizona
Hubertus Strughold, Aerospace Medical Division, USAF, Texas
Harold C. Urey, University of California, San Diego

Appendix III

The Roger Caras Letter

This is the letter that started going out from early January 1966 onwards inquiring as to whether the scientists concerned would consider appearing in the prologue. It was drafted by SK but I think Roger Caras may have changed certain things – for instance, the rather cumbersome styling of titles looks distinctly non-SK. I have re-keyboarded the letter here exactly as it went out and thereby saved it from being littered with *sic*s.

Dear___________

Stanley Kubrick is currently filming a motion picture entitled, "2001: A SPACE ODYSSEY" at M.G.M. Studios in England. The original story for this film was written by Mr. Kubrick and Arthur C. Clarke. It is being shot in Technicolor and will be released world-wide by M. G. M. in Cinerama early in 1967.

The film concerns itself with provocative philosophical and scientific themes. It revolves around the discovery on the Moon thirty-five years from now of the first extraterrestrial artefact. It is determined as the story unfolds that this first evidence represents an intelligence whose origin was other than Earth and that it visited our Moon during the Pleistocene Period.

If you are acquainted with Mr. Kubrick's earlier films ("DR. STRANGELOVE"; "PATHS OF GLORY") and with Mr. Clarke's writings ("CHILDHOOD'S END"; "PROFILES OF THE FUTURE") you will understand that this is not "another science-fiction film", but a very carefully developed extrapolation of what may be the reality of just a few years from now.

Technical advice has been obtained from widely diverse sources such as the National Aeronautics and Space Administration, IBM, Bell Laboratories, Dupont, Minneapolis Honeywell, and nearly forty similar organisations.

As a Foreword to the film, we will have a series of eminent scientists discuss their views on the likelihood of our encountering extraterrestrial intelligence, the possibilities of communications with cultures on other planets, and the probable cultural impact of the first extraterrestrial contact.

These brief interviews, with content determined in advance, can be filmed at the individual's place of work or other suitable setting and should not consume more than an hour of his time.

The purpose of this letter is to seek your participation in this brief introduction. If you can make yourself available to these ends we will arrange a time and place most convenient to your schedule.

Keeping in mind the objectives of this Foreword, to help the audience realise that the basic subject matter of the film is not fantasy but possible if not probable fact, I hope you will be able to favour us with a positive response. The discussions, I must add, will not directly involve the film, but will discuss only related thematic ideas.

Most respectfully,

Roger A. Caras

Hawk Films Ltd.
MGM Studios,
Borehamwood,
Herts,
England.

Historically speaking, this should be two words, Boreham Wood. The affectation was taken up by the local council in the 1960s and promoted as part of, I guess, a 're-branding' of the 'hood.

Appendix IV

Notes on Interviewees

THESE ARE very brief notes and for fuller biographical information the reader should check the standard biographical directories and the Web where details of the subjects' publications will also be found.

ISAAC ASIMOV (1920-92), was born in Russia and taken to the USA at the age of three. Studied chemistry at Columbia University. Biochemist. A prolific science fiction novelist, also a critic and science populariser.

JEREMY BERNSTEIN (1929–), was born in New York. Studied at Harvard. Has been a physicist at Los Alamos and Brookhaven National Laboratories. Staffer at *The New Yorker*. Professor of Physics at the Stevens Institute of Technology since 1967. Wrote a profile of Stanley Kubrick: see Bibliography.

FRANK D. DRAKE (1930–), is a pioneer of radio astronomy and in 1960 conducted Project Ozma at Green Bank, the first programme methodically searching for intelligent extraterrestrial radio signals. He is a professor of Astronomy and Astrophysics at the University of California, Santa Cruz. Formulated the Drake Equation. Interview, pp. 54-85 in David W. Swift's *SETI Pioneers* (1990).

FREDERICK C. DURANT III (1917-2015), B.S. degree in Chemical Engineering from Lehigh University in 1939. Commander, U.S. Navy (retired). Director of Engineering at the Naval Air Rocket Test Station. Former Assistant Director for Astronautics of the National Air and Space Museum. He was the author of one of the earliest scientific investigations into UFOs, convened by the CIA's Office of Scientific Intelligence, in 1953 ('The Durant Report').

FREEMAN J. DYSON (1923–), was born in England and educated at Cambridge. He became a physics professor at Cornell University in 1951. Professor

Emeritus at the Institute for Advanced Studies, Princeton. Proposed the Dyson Sphere. Interview, pp. 311-26 in David W. Swift's *SETI Pioneers* (1990).

GERALD FEINBERG (1933-1992), was chairman of the Columbia University Department of Physics. He was on the board of the Foresight Institute (nanotechnology), took an early interest in cryonics and coined the word 'tachyon' to describe hypothetical faster-than-light particles.

CONSTANTINE D. J. GENERALES JR (1909[?]-1998), was born in Athens, Greece, in 1908 and taken to the States at an early age. He attended universities in Athens, Heidelberg, Paris and Zurich (where he first met and worked with Wernher von Braun, in 1931). Founder and chairman of the Section on Space Medicine of the Medical Society of New York. Co-founder of the New York Academy of Medicine's Section on Biomedical Engineering.

IRVING JOHN GOOD (1916-2009), was born in London. During World War II he worked as a cryptanalyst at Bletchley Park on Ultra under Alan Turing. He was a professor at Trinity College, Oxford, and in the late 1960s went to live and work in the United States.

GERALD HAWKINS (1928-2003), astronomer and archaeo-astronomer, born in Great Yarmouth, England. Gained two doctorates from the University of Manchester and worked with Sir Bernard Lovell. He moved to the States and was Professor of Astronomy and chairman of the Department at Boston University, 1957-69. Twenty years at the Harvard-Smithsonian Observatories in Cambridge, Massachusetts.

FRANCIS J. HEYDEN SJ (1907-91), entered the Society of Jesus in 1924 and was ordained a priest in 1937. He taught astronomy at Georgetown University, Washington DC, 1946-8, and was Professor of Astronomy and Director of the Observatory there, 1948-71. From 1971 until his death he taught physics at Ateneo de Manila and worked at the Quezon City Observatory, Manila.

NORMAN LAMM (1927–), graduated from Yeshiva College, New York, summa cum laude in 1949 and was ordained a rabbi two years later. A noted authority on Jewish philosophy and the author of numerous books. Now Chancellor of Yeshiva after 27 years as President.

SIR BERNARD LOVELL (1913-2012), was a leading British astronomer and radio-astronomer. Professor of Radio Astronomy at Manchester University, 1951-80, and leader of the team that built the Jodrell Bank Radio Telescope, then the largest steerable radio telescope, 1957. He was the telescope's director until 1981.

MARGARET MEAD (1901-78), anthropologist, studied at Columbia University. Widely known for her 1929 study, *Coming of Age in Samoa*. Was for many years at the American Museum of Natural History in New York City.

MARVIN MINSKY (1927-2016), was born in New York. Toshiba Professor of Media Arts and Sciences and Professor of Electrical Engineering and Computer Science at the Massachusetts Institute of Technology. A pioneer of intelligence-based mechanical robotics and the development of artificial intelligence. Interview, pp. 15-31, in David G. Stork's *HAL's Legacy: 2001's Computer as Dream and Reality* (1997).

ORMOND G. MITCHELL was Professor Emeritus in the Department of Anatomy at the New York University Dental Center. Bachelor's degree in biology from San Diego University in 1948. A technical adviser on *2001: A Space Odyssey*. He died in Beverly Hills in 2001.

PHILIP MORRISON (1915–), worked on the Manhattan Project, 1942-6, then joined the physics department at Cornell. Since 1965 he has been at MIT. In 1959 with Giuseppe Cocconi he published an important and influential paper arguing for a search for extraterrestrial signals. Interview, pp. 19-48 in David W. Swift's *SETI Pioneers* (1990).

A. I. OPARIN (1894-1980), Russian biochemist and botanist, studied plant physiology at Moscow State University. In the 1920s he developed the first modern theory as to the origin of life, arguing that it arose in a primeval 'soup' of bio-molecules (the basis for Urey and Miller's experiments in 1953).

HARLOW SHAPLEY (1885-1972), astrophysicist, educator, author, worked at the Mount Wilson Observatory and subsequently became Director of the Harvard University Observatory, 1921-52. His greatest single contribution to science was discovering the dimensions of our galaxy and the location of its centre.

S. M. SIEGEL (1928-90), born Kansas City. Educated Chicago University. Botanist and expert on plant chemistry specialising in plant peroxidases. Professor, later chairman of the Botany Department, University of Hawaii in Honolulu, 1967-90.

B. F. SKINNER (1904-90), born in Pennsylvania. Leading behavioural psychologist who reached a wide public audience through his controversial book, *Walden Two* (1948). From 1948 until his retirement he was at Harvard.

FRED L. WHIPPLE (1906-2004), born in Iowa, was the most acclaimed American astronomer of the twentieth century. Studied in California and became Professor of Astronomy at Harvard in 1945. Principally known for his work on the solar system and comets.

Afterword

by Frederick I. Ordway III

ON 1 FEBRUARY, shortly after 9am, Eastern Standard Time, I sat down in front of my PC. I'd promised volume editor Tony Frewin that I'd prepare an afterword before the middle of the month.

So why not get started early?

I'd jotted down some notes when my wife called from upstairs. 'Something's gone wrong with the shuttle,' she said – her radio was on, mine wasn't. By shuttle I knew she meant the returning Columbia on the STS-107 all-science mission.

I rushed to tune in CNN and NASA-TV where I learned that communications with Columbia had ceased just before 9am Eastern Time and that it had begun to break up over north-central Texas 15 minutes before scheduled touchdown at the Kennedy Space Centre. This was the second shuttle tragedy, almost 17 years to the day after Challenger had exploded 73 seconds following launch on 28 January 1986. And, coincidentally, 36 years and a few days after three astronauts had died during a ground test of the Apollo 1 command module (27 January 1967). I knew that once again there would be delays in America's space programme.

My mind flashed back to January 1965 when I joined Stanley Kubrick and Arthur C. Clarke as the film's technical consultant. By that time, the four-mission, one-man Mercury satellite programme (1961-63) had been completed and the two-man, ten-mission Gemini programme (1965-66) was getting under way. We were then about half way along the road towards undertaking manned Apollo expeditions to the Moon (1968-72). In mid-May 1973, the Skylab space station followed, which was visited by three different astronaut crews during its active lifetime.

In the unmanned realm, by the time *2001: A Space Odyssey* pre-production had begun, the Army had sent Pioneer 4 past the Moon and into orbit around the Sun (1959) and three years later NASA's Mariner 2 flew by Venus, the world's first successful planetary probe. Then, in mid-1964, NASA's Ranger 7 hit the lunar surface, returning over 4,000 photos before crashing. The Soviets were also moving ahead with their space programme.

In those days, the future of spaceflight indeed looked bright. And the year

2001 was still more than 35 years in the future. Almost anything seemed possible.

Also colouring our optimism at the time were NASA-sponsored studies that called for manned missions to the nearer planets in the relatively near future. For example, in Huntsville, Alabama, my home at the time and where I had worked since the middle 1950s with the Wernher von Braun 'rocket team,' a project known as EMPIRE (for Early Manned Planetary-Interplanetary Round-trip Expeditions) was under way at the NASA-Marshall Space Flight Center's Future Projects Office. With the support of three industrial contractors, engineers were examining the use of heavy-lift launch vehicles (upgraded Saturn Vs or perhaps larger Novas) to lift into Earth orbit the components from which nuclear-powered interplanetary spaceships would be assembled. Among the missions to be executed *by the mid-1970*s: manned Venus and Mars flybys, Venus and Mars orbital capture, and possibly even a touchdown on the Martian surface.

Small wonder, then, our exhilaration as we began laying out an array of vehicular and lunar-base concepts that we imagined might exist in the year 2001.

For starters, we visualised a fully reusable, commercially operated Earth-to-orbit space shuttle named Orion III that was to fly almost as routinely as today's airliners – and by Pan American. That distinguished airline company later collapsed.

In reality, by the year 2001 and continuing to the present, we have relied on experimental space shuttles, operated by the government, and used to carry crews, construction experts and pay-load specialists into orbit. Between April 1981 and February 2003, the shuttle manifest has been marred by two accidents in 113 flights – Challenger and Columbia.

By 2001, no large, rotating station typified by *2001*'s Space Station V was in orbit, and it still isn't. But the far less ambitious International Space Station continues under construction and hopefully will be completed during the present first decade. Regretfully, it won't enjoy the features of Space Station V, but some later station might. For example, in the film we see visiting personnel undergo voice-print identification upon entering the rim section. Such systems do exist today and are available commercially, but not on the International Space Station.

Picture-phones or vision-phones providing two-way audio and visual communications also exist but are used primarily for business video teleconferencing here on Earth. In fact, when the American Telephone and Telegraph Company demonstrated the first commercial Picture-phone in Pittsburgh in 1970, the company got more press attention than paying-customer interest.

Since then, the idea just hasn't caught on with the general public, though it may in the years ahead. One can, of course, attach digital cameras to PCs and send live video, but the practice is not yet common. And certainly not on a space station.

Hilton and Howard Johnson, featured commercial tenants aboard Space Station V, are still around, but not in orbit. The advent of space tourism may lead to some of the services we took for granted in 2001. But the future is still cloudy in this area.

The Aries 1B, 2001's Earth-orbit-to-lunar-surface shuttle has no counterpart today, except on the drawing board. Back in the late 1960s and early 1970s, however, plans were pursued that would have led by the mid-1980s to nuclear-powered space vehicles shuttling back and forth between Earth orbit and lunar orbit. The United States nuclear rocket programme was cancelled in 1972, and today there are no active, funded plans for such craft nor for lunar orbital stations and surface bases. *2001*'s Clavius-base remains a dream. But sporadic base and outpost studies continue by return-to-the-Moon devotees.

For across-the-Moon travel, we chose what we called the lunar or rocket bus, hardly a more romantic term than shuttle. There's no telling when such a vehicle will appear in reality. In 1965, we failed to anticipate the demise of manned lunar activities that would follow the completion of the Apollo programme. It just didn't occur to us at the time that the United States would back away from then still-to-be-achieved lunar triumphs and content itself with circling the Earth during the decades that followed.

After the lunar episodes, *2001*'s action shifts to the expedition to Jupiter and life aboard the spaceship Discovery. No such ship exists today except in the imaginations of designers and engineers. Sad, perhaps, but since 2001 unmanned spacecraft carrying such names as Mariner, Pioneer, Viking, Voyager, Magellan, Galileo and Cassini (still en route to Saturn) have remotely explored all of the planets but Pluto, and most of their moons.

We now come to what many believe to be Discovery's most memorable entity: the HAL 9000 computer. How does 'he' stand up as we approach four decades since *2001* pre-production began? Extremely fast supercomputers, parallel processing and global networks as well as powerful graphical displays are with us today. But not the voice-output capabilities of HAL, who speaks, listens, sees, appears to reason and may exhibit emotions (such as self-doubt). According to cognitive scientist Donald A. Norman, 'we are still a very long way from creating a system as powerful as HAL. A very long way.'

Stephen Wolfram, the creator of the Mathematica firm, several years ago suggested that the reason why a HAL-like machine doesn't exist yet is 'basically just an accident.' The right person to lead the way to a HAL simply didn't 'come along' in time. 'When there are machines that think,' he mused, 'are we going to look back and say, "Why, all of this could definitely have been done . . . in the 90s, perhaps even in the 80s! How on earth did this get missed?"'

HAL talked (through the voice of Canadian Shakespearian actor Douglas Rain). Such vocal perfection does not yet exist, though a Bell Laboratories computer can read aloud in English and a few other languages. But the voice sounds like a computer, not Douglas Rain. Furthermore, we are nowhere near developing a system that understands what it is saying.

Voice- or speech-recognition technology allows modern computers to recognise spoken words provided they are enunciated clearly and individually – or, at most, in short phrases. And by a voice known to the computer. More simplistically, computers can recognise, to a limited extent, words uttered by the average person in response to specific menu commands. A number of companies provide speech-recognition software used by modern PCs with vocabularies up to 60,000 words.

Computer-vision technology has now reached the point where documents can be read if the surface is flat and motionless. Computers are poor at making out three-dimensional objects that move around, though research in this area is progressing.

A key story point in the movie was HAL's ability to read lips as Frank Poole and David Bowman talked, privately, in the Space Pod. Some work is being undertaken to assist computers to respond to speech with, according to machine-learning and perception researcher David G. Stork, 'encouraging' results. He reckons that computer speech-reading may be in the development state comparable to that of acoustic recognition in the 1940s.

Running a spaceship by computer became routine with the advent of application software. Even computer self-diagnosis is with us today. A case in point is Deep Space 1, the first spacecraft in NASA's New Millennium Program, which was designed to test and validate new technologies, one being an autonomous navigation system or AutoNav. In brief, it allowed the spacecraft to take over parts of an interplanetary navigation mission that ordinarily would be the task of ground controllers. Deep Space 1 could find its location in the solar system by imaging known asteroids and comparing their positions to background stars. (Stored in the spacecraft's computer memory were the orbits of 250 asteroids and the positions of 250,000 stars.) Trajectory changes were governed by changes in the profile of the craft's ion propulsion system.

Another example is the on-board fault-protection used by the Saturn-bound Cassini spacecraft, which is due to reach its target in July 2004. If a substantial fault is discovered, on-board computers automatically initiate a variety of 'saving' actions without any intervention from controllers back on Earth. Modern space-flight would be impossible without computers.

So far, reasoning, emotions and common sense are attributes that have not been demonstrated in computers. According to Dr. Marvin Minsky of the Massachusetts Institute of Technology, 'Common sense is knowing maybe 30 or 50 million things about the world and having them represented so that when something happens, you can make analogies with others.' Computers cannot yet do this. Studies have been made regarding recognition and expression of emotions but results have been nebulous at best. As for emotions, Minsky thinks that 'you're going to have to go to great lengths to prevent advanced computers from having some acquisitiveness and the need to control things. Because to solve a problem, you have to have the resources…'

This brings us square into the debate regarding artificial intelligence which, despite considerable optimism back in the mid-1960s when we were working on *2001*, remains an incredibly difficult branch of human endeavour. Involved are not only reasoning, the acquisition of prodigious amounts of information, and the ability to create and synthesise new knowledge, but judgement; the making of rational decisions; and reliance on plain common sense. A possible road to AI is the reverse engineering of the human brain. To mathematician Sir Roger Penrose of Oxford University, 'Understanding is outside a computer. It doesn't understand.'

Well, for the time being that may be. And for the time being, progress towards the manned exploration of the solar system may seem agonisingly slow to fans of *2001: A Space Odyssey* and other intellectual and visual projections into the future.

But spaceflight will continue and the pauses and delays that sometimes frustrate us will be relegated to footnotes in the long stretch of human history. Meanwhile, as we seek out the secrets of Jupiter and other worlds in the solar system, we will at the same time be learning more about the stars in our region of the Milky Way galaxy. Already a post-Hubble space telescope named after NASA Apollo-era administrator James Webb is under development, and NASA is planning space interferometry and terrestrial planet finder missions for the years ahead. Using such technology, we'll reach the point where Earth-like extra-solar planets will be detected and examined. Now that over a hundred giant planets have been discovered circling other stars, the chances are increasingly favourable that small ones exist, too.

And perhaps one day, intelligent signals will be received from one or more of them, or from more distant sources. If and when that happens, we may link up with an extrasolar or even a galactic network, one of the premises of *2001: A Space Odyssey*.

During an interview at the NASA-Johnson Space Center in Houston, Texas on 19 September 2001, first-man-on-the-Moon astronaut Neil A. Armstrong had this to say when asked about space and the movies: 'I thought *2001*, which was many decades ago now, was a very fine film, very authentic in terms of the way space looks and the way vehicles move and trajectories, and so on. Many of the more recent space fiction movies have much less realism than *2001* did. They're more exciting, but not realistic.'

Washington DC, 2003

Arthur C. Clarke, Keir Dullea and SK.
Probably at *2001*'s New York premiere in April 1968

Bibliography & Web Resources

THIS BIBLIOGRAPHY is divided into four sections: SK's seven key texts (as discussed in the Introduction); general works on extraterrestrial intelligence; *2001: A Space Odyssey*, the film; the cinema of SK.

I think one may reasonably claim that more books and articles on extraterrestrial life have been published in the last forty years than in the preceding two thousand years. The books listed below are all to be recommended in one way or another.

There are three bibliographies of the subject: *Extraterrestrial Life: A Bibliography, Part I, Report Literature, 1952-1964* (NASA, 1964) and *Part II, Published Literature, 1900-1964* (NASA, 1965) and, updating to the late 1970s, *A Bibliography on the Search for Extraterrestrial Intelligence* (NASA, 1978) by Eugene F. Mallove, Mary M. Connors, Robert L. Forward and Zbigniew Paprotny. The references in the post-1970s volumes below will give any interested reader a good idea of the range and extent of the subsequent literature.

I have eschewed the subject of UFOs – a minefield of mis- and disinformation – principally because of the anecdotal rather than scientific approach that largely characterises the subject. The field is polarised between those who will believe just about anything, even abductions (Whitley Strieber, say), and those who won't believe anything (Philip J. Klass, say). A middle ground is hard to find. However, if pushed to recommend some titles I would opt for three works. First, the collection of papers edited by Carl Sagan and Thornton Page, *UFO's – A Scientific Debate* (New York: W. W. Norton, 1972) which, though published some thirty years ago, nonetheless has criticisms and observations that hold good today. Second, David M. Jacobs' study of *The UFO Controversy in America* (Bloomington: Indiana University Press, 1975), and third, Randall Fitzgerald's *Cosmic Test Tube: Extraterrestrial Contact, Theories & Evidence* (Los Angeles: Moon Lake, 1998), essentially a fulsome annotated bibliography of over 200 works covering not only UFOs but also 'ancient astronauts,' contactees and abductees, debunkers, and SETI. A useful guidebook to the area.

It must have been by the early 1970s that the number of books about Stanley Kubrick far exceeded the number of films he had actually made. Every year produces more titles and, like books on other film directors, most of them are either scrapbooks of facts and stills or meandering critical interpretations. Cinema criticism has produced very few fine critics. However, the titles below are all recommended. Also recommended is the documentary made by Jan Harlan, SK's long-time associate producer, *Stanley Kubrick: A Life in Pictures* (Warner Bros, 2001, 150 minutes). This is currently available on video and DVD and presents a history and portrait by those who knew him.

A] EXTRATERRESTRIAL INTELLIGENCE: STANLEY KUBRICK'S SEVEN KEY TEXTS:

CAMERON, A. G. W. (Ed.). *Interstellar Communication: The Search for Extraterrestrial Life.* New York: W. A. Benjamin, 1963.

CLARKE, ARTHUR C. *Profiles of the Future: An Inquiry into the Limits of the Possible.* New York: Harper & Row, 1962.

GOOD, I. J. (Ed.). *The Scientist Speculates: An Anthology of Partly-Baked Ideas.* New York: Basic Books, 1962.

MACGOWAN, ROGER A. & ORDWAY III, FREDERICK I. *Intelligence in the Universe.* Englewood Cliffs, New Jersey: Prentice-Hall, 1966.

SAGAN, CARL, & SHKLOVSKII, I. S. *Intelligent Life in the Universe.* San Francisco: Holden-Day, 1966.

SHAPLEY, HARLOW. *The View from a Distant Star: Man's Future in the Universe.* New York: Basic Books, 1963.

SULLIVAN, WALTER. *We Are Not Alone: The Search for Intelligent Life on Other Worlds.* New York, 1964.

B] EXTRATERRESTRIAL INTELLIGENCE: GENERAL WORKS:

ACZEL, AMIR. *Probability 1: Why There Must be Intelligent Life in the Universe.* London: Little, Brown, 1998. A statistical probability study keying off from the Drake Equation.

ASHPOLE, EDWARD. *The Search for Extraterrestrial Intelligence.* London: Blandford Press, 1989. A good introductory study.

ASIMOV, ISAAC: *Is Anyone There?* New York: Doubleday, 1967.
Popularly written essays on all aspects of the subject with the author's characteristic wit and learning.

BILLINGHAM, JOHN (Ed.). *Life in the Universe.* Cambridge, Mass.: MIT Press, 1981. Some twenty-nine papers, largely technical, delivered at the Conference on Life in the Universe held at the NASA Ames Research Center in 1979. Includes contributions by Ronald Bracewell, Frank D. Drake and

Bernard M. Oliver. Billingham was Chief of the Extraterrestrial Research Division at Ames.

BOVA, BEN & PREISS, BYRON (Eds.). *First Contact: The Search for Extraterrestrial Intelligence.* London: Headline, 1990. Twenty-two essays including Frank Drake's 'The Drake Equation: A Reappraisal,' Philip Morrison's 'Reflections on the Bigger Picture,' and Arthur C. Clarke's 'Where Are They?'

BRACEWELL, RONALD N. *The Galactic Club: Intelligent Life in Outer Space.* New York: W. W. Norton, 1974. A wide-ranging and key work by the famous radio astronomer. A full discussion of what became known as 'Bracewell Probes.'

CHRISTIAN, JAMES L. (Ed.). *Extraterrestrial Intelligence: The First Encounter.* Buffalo, New York: Prometheus Books, 1976. Fourteen essays by divers hands including Ray Bradbury and Isaac Asimov.

CROWE, MICHAEL J. *The Extraterrestrial Life Debate, 1750-1900.* Cambridge: Cambridge University Press, 1986. New edition with a new preface: New York, Dover Books, 1999. The extraterrestrial life debate was earlier known as the 'plurality of worlds' debate and Crowe's study is unlikely ever to be surpassed. Enormously rich and detailed. Dick's volume below may be considered a sequel.

DARLING, DAVID. *The Extraterrestrial Encyclopedia: An Alphabetical Reference to All Life in the Universe.* New York: Three Rivers Press, 2000. This should be on the shelf of everyone interested in extraterrestrial intelligence. Essential. See Web Resources below for Darling's website. *Life Everywhere: The Maverick Science of Astrobiology.* New York: Perseus Books, 2001. An excellent up-to-date introduction.

DAVIES, PAUL. *Are We Alone? Philosophical Implications of the Discovery of Extraterrestrial Life.* New York: Basic Books, 1995. Discusses areas frequently overlooked in scientific-based studies: the philosophical, theological and political implications.

DICK, STEVEN J. *The Biological Universe: The Twentieth-Century Extraterrestrial Life Debate and the Limits of Science.* Cambridge: Cambridge University Press, 1996. A magisterial and exhaustive history of the subject in the last century and immensely readable. Dick has read everything, spoken to everyone. A great work of scholarship and a sequel, as it were, to Crowe's work, above. See also Swift, below.

DRAKE, FRANK & SOBEL, DAVA. *Is Anyone Out There? The Scientific Search for Extraterrestrial Intelligence.* London: Souvenir Press, 1992. A first-hand account by Drake.

FEINBERG, GERALD & SHAPIRO, ROBERT. *Life Beyond Earth: The*

Intelligent Earthling's Guide to Life in the Universe. New York: William Morrow, 1980. Perceptive study of the origins and diversity of life that argues convincingly for its prevalence throughout the universe but not necessarily in the carbon-based forms known to us.

GOLDSMITH, DONALD, & OWEN, TOBIAS. *The Search for Life in the Universe.* Reading, Mass.: Addison-Wesley, 1993. Comprehensive and illustrated throughout. A standard text.

HARRRISON, ALBERT A. *After Contact: The Human Response to Extraterrestrial Life.* New York: Perseus Publishing, 2002. Comprehensive and pentrating. Not an introductory volume. Reflective and wide ranging.

HART, MICHAEL F. & ZUCKERMAN, BEN. *Extraterrestrials: Where are They?* Cambridge: Cambridge University Press, 1995 (2nd ed). Twenty-two papers, including contributions from Ronald Bracewell, Jill Tarter and Freeman Dyson. Not an introductory work. An important collection.

HEIDMANN, JEAN. *Extraterrestrial Intelligence.* Cambridge: Cambridge University Press, 1995. A very comprehensive yet concise introduction to all aspects of the subject. Some technical material, but explained in terms the layperson would understand.

KOERNER, DAVID & LEVAY, SIMON. *Here Be Dragons: The Scientific Quest for Extraterrestrial Life.* Oxford and New York: Oxford University Press, 2000. A good contemporary account of SETI's state-of-play. Interviews with scientists in the field (including, just to redress the balance, Duane Gish of the Institute for Creation Research in San Diego who is attempting to scientifically prove that God made the universe in a week).

PARKER, BARRY. *Alien Life: The Search for Extraterrestrials and Beyond.* Reading, Mass.: Perseus, 1998. Comprehensive chapters on Mars and the Mars meteorite. Illustrated.

PUCCETTI, ROLAND. *Persons: A Study of Possible Moral Agents in the Universe.* London: Macmillan, 1968. A philosophical inquiry into the concept of a 'person.' Puccetti examines the supercomputer, the organic artefact and extraterrestrials. Challenging and perceptive.

REGIS, EDWARD, Jr. (Ed.). *Extraterrestrials: Science and Alien Intelligence.* Cambridge: Cambridge University Press, 1985. An important collection of essays including the paleobiolgist David M. Raup's 'ETI without Intelligence' which argues that the fossil record shows life may have originated independently on Earth a number of times. Also Frank J. Tipler's 'Extraterrestrial Intelligent Beings Do Not Exist' and the reply by Carl Sagan and William I. Newman, 'The Solipsist Approach to Extraterrestrial Intelligence.'

SAGAN, CARL (Ed.). *Communication with Extraterrestrial Intelligence* (CETI).

Cambridge, Mass: MIT Press, 1973. Papers and discussions from the very first international conference on extraterestrial life, held in September 1971 in Soviet Armenia. Contributions from I. S. Shklovskii, Frank D. Drake, Philip Morrison, Freeman Dyson, Marvin Minsky, Francis Crick and others.

The Cosmic Connection: An Extraterrestrial Perspective. New York: Anchor Press/Doubleday, 1973. Written for general readership in non-technical terms.

SHAPIRO, ROBERT. *Planetary Dreams: The Quest to Discover Life Beyond Earth.* New York: Wiley, 1999. Lively and comprehensive. One of the best of the more recent additions to the subject.

SHOSTAK, SETH. *Sharing the Universe: Perspectives on Extraterrestrial Life.* Berkeley, California: Berkeley Hills Books, 1998. Excellent introductory volume by the Public Programs Scientist of the SETI Institute in California.

SWIFT, DAVID W. *SETI Pioneers: Scientists Talk about Their Search for Extraterrestrial Intelligence.* Tucson: University of Arizona Press, 1990. Extended interviews with sixteen scientists who are or were at the centre of SETI research: John Billingham, Ronald Bracewell, Melvin Calvin, Giuseppe Cocconi, Frank Drake, Freeman Dyson, Nikolai Kardashev, John Kraus, Philip Morrison, Bernard Oliver, Carl Sagan, Kunitomo Sakurai, Charles Seeger, Iosef Shklovskii, Jill Cornell Tarter, and Vasevolod Troitskii. An essential companion to Dick, above.

C] THE FILM: *2001: A SPACE ODYSSEY*:

AGEL, JEROME. *The Making of Kubrick's 2001.* New York: Signet/New American Library, 1970. A grab-bag of articles, letters, reviews, interviews and associated material relating to the movie (it even has the parody from *Mad* magazine). Includes the September 1968 *Playboy* interview with SK, Arthur C. Clarke's original short story, 'The Sentinel,' Jeremy Bernstein's profile of SK from *The New Yorker*, and an extended illustrated section. Excerpts from the interviews with the scientists were also included.

BERNSTEIN, JEREMY. *A Comprehensible World: On Modern Science and Its Origins.* New York: Random House, 1967. Contains a tribute to Arthur C. Clarke and a profile of SK that first appeared in *The New Yorker.*

BIZONY, PIERS. *2001: Filming the Future.* London: Aurum Press, 1994 (new revised edition, 2000). Primarily a study of the making of the film, the nuts and bolts. Well illustrated throughout.

CLARKE, ARTHUR C. *2001: A Space Odyssey* (a novel by). London: Hutchinson, 1968. *The Lost Worlds of 2001*. New York: New American Library, 1972. Clarke's account of the genesis of *2001* and his subsequent collaboration with SK. *Greetings, Carbon-Based Bipeds! A Vision of the 20th Century as it Happened* (Ian T. Macauley, ed.). London: Harper-Collins, 1999. A Clarke chrestomathy with much about *2001* and SK.

GEDULD, CAROLYN. *Filmguide to 2001: A Space Odyssey*. Bloomington: Indiana University Press, 1973. SK always considered this one of the more intelligent analyses.

RICHTER, DAN. *Moonwatcher's Memoir: A Diary of 2001: A Space Odyssey*. New York: Carrol & Graf, 2002. Richter played Moonwatcher and choreographed the Dawn of Man sequence. An illuminating account of the day-to-day moviemaking on the picture.

SCHWAM, STEPHANIE. *The Making of 2001*: A *Space Odyssey*. New York: The Modern Library, 2000. (The Modern Library 'Movies' Series: Series Editor, Martin Scorsese). A collection of essays and articles that overlaps Agel, above.

STORK, DAVID G. *HAL's Legacy: 2001's Computer as Dream and Reality*. Cambridge, Mass.: MIT Press, 1997. Several scientists explore the feasibility of achieving a real life HAL, science fiction's most famous computer, and discuss related topics and prospects in artificial intelligence.

D] THE CINEMA OF STANLEY KUBRICK:

CIMENT, MICHEL. *Kubrick: The Definitive Edition*. New York: Farrar, Straus and Giroux, 2001. A new, updated edition of the volume first published in 1983. Ciment, a leading French critic, has written a detailed and perceptive account of the films. A wealth of illustrations intelligently chosen. Interviews with collaborators. A key work.

COYLE, WALLACE. *Stanley Kubrick: A Guide to References and Resources*. Boston, Mass.: G. K. Hall, 1980. Detailed and exhaustive up until the date of publication. Full credit and cast lists for the films and, amongst much else, an 80-page bibliography of articles, reviews, books, etc. For an updating of the bibliography and references see Gene Phillips' article on SK in *The International Dictionary of Films and Filmmakers* (Detroit: St. James Press, 1999 [4th edn]) and the Phillips/Hill Encyclopedia, below.

HERR, MICHAEL. *Stanley Kubrick*. New York: Grove/Atlantic Press, 2000. An affectionate and personal memoir.

KUBRICK, CHRISTIANE. *Stanley Kubrick: A Life in Pictures*. London: Little,

Brown, 2002. A pictorial biography of SK compiled and written by his wife and based on the family archive. Nearly all of the photographs have never been published before. The appendix contains the most exhaustive cast and credits list of the films, frequently with restored names.

LOBRUTTO, VINCENT. *Stanley Kubrick: A Biography. New York*: Donald I. Fine/Penguin Books, 1997. A highly detailed work that is unlikely to be surpassed.

PHILLIPS, GENE D. *Stanley Kubrick: A Film Odyssey*. New York: Big Apple/ Popular Library Film Series, 1977. *Major Film Directors of the American and British Cinema*. Bethlehem, Pennsylvania: Lehigh University Press, 1999 (rev. edn). *Stanley Kubrick: Interviews*. Jackson: University Press of Mississippi, 2001. With Rodney Hill. *The Encyclopedia of Stanley Kubrick*. New York: Checkmark Books/Facts on File, 2002. SK thought highly of Gene Phillips as a critic and gave him every assistance on the first two works. Essential reading. As indeed is the Interviews volume that gathers together some sixteen interviews that span SK's career. The *Encyclopedia* is a good first port of call for any serious study of SK's works.

WALKER, ALEXANDER. *Stanley Kubrick, Director*. New York: W. W. Norton, 1999 (with 'A Visual Analysis by Sybil Taylor and Ulrich Ruchti') Walker knew SK for nearly forty years and writes perceptively of the movies.

WEB RESOURCES

NEEDLESS TO SAY, all Websites detailed here are prefixed by http://
The best starting point for information on Stanley Kubrick is the official SK Warner Bros site:

www.kubrickfilms.com.

More links than you would have thought possible are at:

www.geocities.com/SunsetStrip/Studio/5139/kubrick.html

A search on Google or any other good search engine will produce even more. 2001 itself has spawned many sites, amongst which:

www.palantir.net/2001/links.html
www.underview.com/2001.html

www.filmsite.org/twot.html

Christiane Kubrick's website is to be found at:

www.christianekubrick.net.

David Darling, the begetter of *The Extraterrestrial Encyclopedia* (2000), has a rich and well designed website that is maintained and updated regularly and one that should be on everybody's cyberspace itinerary. It contains links to all major sites concerned with life in space:

www.angelfire.com/on2/david darling

Below are the main SETI and SETI-related websites.

Drake Equation www.station1.net/DouglasJones/drake.html
NASA Astrobiology Institute www.nai.arc.nasa.gov
NASA Exobiology Branch www.exobiology.nasa.gov
SETI@Home www.setiathome.ssl.berkeley.edu/
SETI Institute www.seti.org
SETI League www.setileague.org

L'Envoi

A Man said to the universe:
'Sir, I exist!'
'However,' replied the universe,
'The fact has not created in me
A sense of obligation.'

STEPHEN CRANE *War is Kind* (1899)

Published in the U.K. by

Ashgrove Publishing

an imprint of:

Hollydata Publishers Ltd
21 Ewen House
Caledonian Road
London
N1 0SH

ISBN 978 185398 175 3

Reprinted 2026